Field Guide to the Primates of Indonesia

Field Guide
to the Primates
of Indonesia

Jatna Supriatna

 Springer

Jatna Supriatna
Department of Biology
University of Indonesia
Depok, Indonesia

ISBN 978-3-030-83205-6   ISBN 978-3-030-83206-3   (eBook)
https://doi.org/10.1007/978-3-030-83206-3

Cover illustration: Book Cover created already by OBOR

This Springer imprint is published by the registered company Springer Nature Switzerland AG
The registered company address is: Gewerbestrasse 11, 6330 Cham, Switzerland

FOREWORD

It gives me great pleasure to write this foreword to the second edition of the Primates of Indonesia by my very good friend and long-time colleague Pak Jatna Supriatna. This book is especially important because Indonesia ranks, with Brazil, as one of the world's top two megadiversity countries, and is also among the top four in global primate diversity, accompanied by Brazil, Madagascar, and the Democratic Republic of Congo. With five families, 11 genera, 63 species, and 81 taxa of nonhuman primates, Indonesia is third, behind only Brazil and Madagascar. Primate endemism is also extremely high, with two genera (Tarsius and Simias), 42 species and 61 taxa of primates found nowhere else on Earth, numbers exceeded only by Brazil and Madagascar.

We continue to learn more about Indonesia's primates every year, with many treasures still to be discovered. Three primates endemic to Indonesia were described for the first time just in the past year. One was the Tapanuli Orangutan (*Pongo tapanuliensis*) on the island of Sumatra, the first new species of great ape discovered since the Bonobo in 1929, an event of truly global and historic interest. In addition, two new tarsiers were described, both from Sulawesi. One, the Manado Spectral Tarsier (*Tarsius spectrumgurskyae*), was named after long-time tarsier researcher Dr. Sharon Gursky, and the other, the Bumbulan or Supriatna's Tarsier (*Tarsius supriatnai*), was named after none other than the author of this book.

I first met Jatna Supriatna more than 35 years ago when he was a graduate student at the University of New Mexico, studying one of my favorite Neotropical genera, the spider monkeys of the genus Ateles. Pak Jatna was one of the first Indonesian members of our IUCN/SSC Primate Specialist Group, and he has served as Vice-Chair for Southeast Asia in that group since November 1994. Later, I got to know him even better on the occasion of the XV Congress of the International Primatological Society, held in Bali in 1994. Right after that very successful event, Pak Jatna and I, together with my then wife Cristina and my two sons John and Mickey, made an amazing trip to the island of Sulawesi, and especially Togian Islands, where we had the good fortune to observe the Togian Island Macaque. At that time, I tried to recruit Pak Jatna to come and work for Conservation International as our Indonesia Program Director, and was eventually successful.

This was truly historic since it marked the first time that a major international conservation organization hired an Indonesian national to head its program in that country. He did an outstanding job in that role from 1994 to 2009, first as Executive Director and then as Vice-President, and created many new programs and worked to save many critical ecosystems and species. During his tenure as leader of CI's program we did a number of other trips together, the most memorable of which was in 1999 when we went to Siberut in the Mentawai Islands, one of the world's highest primate conservation priorities. This amazing trip also included Dr. Gustavo Fonseca, today Director of Programs at the Global Environment Facility, and, again, my son John, now considerably older and well on the way to becoming an accomplished ornithologist. Over the years, Pak Jatna has been a close friend, not only to me but to my family as well. Indeed, one of the highlights of my career was to be able to present his tarsier species to him at a Conservation International event in Bali in March, 2017. We managed to keep secret the fact that we were naming the species in his honor up until that very moment, and I could tell immediately from his response how important this was to him.

Today, Jatna occupies so many key leadership positions that it is difficult to keep track of them all, but he is without doubt his country's greatest living biodiversity expert. This new book on the primates of Indonesia is a follow-up to the first edition he published with Edy Hendras Wahyono in 2000 and with Rizki Ramadhan in 2016. Needless to say, it is greatly expanded and updated, and an outstanding piece of work. This time around, it is being published in both Bahasa Indonesia and English, which means that it will have a global impact, while at the same time serving its principal role as a vital resource for the younger generation of Indonesian primatologists and conservationists.

Congratulations to Pak Jatna Supriatna for yet another major contribution to our understanding of Indonesian biodiversity, promoting the conservation of the wildlife and ecosystems in his very special country.

Russell A. Mittermeier, Ph.D.
Chief Conservation Officer,
Global Wildlife Conservation, and Chair, IUCN/SSC
Primate Specialist Group

FOREWORD

I really welcome this new book "Field Guide to the Primates of Indonesia". This book has been waited for such a long time by many scientists in Indonesia and also abroad. This book will be important for students, scientists, tourists who want to know about Indonesia's primate in the forest and also in the parks in Indonesia. It is very important for my rangers all over Indonesia to know each primate species, its behavior, geographical boundaries and how to protect them. Some of these primates have been categorized as highly critically endangered by IUCN due to their small populations such as Tapanuli and Sumatra orangutans, being pet lovers such as gibbons, macaques. leaf monkeys or many of them hunted for food by local communities. Some have been traded globally such as slow loris while others are pictured to be crop raiders such as long tail macaques. This crop raider means becoming a pest for farmers who live near the jungle. But many of Indonesia's primates become tourist attractions. Many tourists want to see orangutans in Tanjung Puting National park in Central Kalimantan and Gunung Leuser National Parks in Aceh and some want to see the smallest primate in Indonesia, the tarsiers who live in Tangkoko Batuangus Recreation Park in North Sulawesi.

I have been impressed with Prof. Jatna Supriatna who has written many books and papers on Indonesia's biodiversity while he is also actively teaching at the University of Indonesia and doing his filed research plus helping many NGOs develop conservation programs in many sites. His perseverance and endurance should be appreciated by all of us and his co-workers. I had worked side by side with him at Conservation International Indonesia a decade ago and so I have known his dedication to help conserve and develop our biodiversity. I hope he still has time to write more books and papers on Indonesia's biodiversity uniqueness and richness, so many of us will benefit from his hard work

Ir. Wiratno M.Sc
Director General of Natural Resources and Ecosystem
Min of Environment and Forestry of Republic of Indonesia

PREFACE

In 2000, my colleague Edy Hendras and I wrote a pocket-sized guide book to the primates of Indonesia. The book was well received as it was the first to describe the behavior, ecology, distribution, conservation status, and where in the field to find. all of the 44 species of Indonesian primates recognized at that time. Since the book was written in Bahasa Indonesia, there have been many requests to translate it into English. However, that translation has been pending for a long-time due to my focus on other activities.

One of those other activities was the idea of a more scientific publication on Indonesian primates, conceived in collaboration with Dr. Sharon Gursky, a colleague from my graduate school days at the University of New Mexico who is now teaching at Texas A & M University. Together, we convened 38 scientists from Indonesia and abroad who had studied primates in the wild in Indonesia, and invited them to contribute to a book describing the evolutionary history, ecology, behavior, distributional ranges, and conservation status of Indonesia's primates. This book, entitled "Indonesian Primates" was published in 2010 by Springer. Following this, my colleague Rizki Ramadhan and I published "Primate Tourism in Indonesia". By this time, a major revision of Asian Primates (Roos et al., 2014) had recognised 59 Indonesian species so this new book described those 59 plus several new species described. I also add information on where they could be found and how to get there. It was published in 2016 in the Indonesian language by Obor Foundation in Jakarta.

But a translation of the original guide book was still pending and with the recent taxonomic revision, it seemed more appropriate to write a new book, rather than simply translate the original. Hence, this book was conceived. As a field guide, it is intended to be as accessible as possible to tourists visiting the Indonesian jungle who want to get a better understanding of the ecology, behavior and conservation status of the species, as well as observe them in the wild. It should also be of use to students studying primatology. However, the main objective is to provide a means for the lay person to become acquainted with the primates and their habitats, and thereby foster an interest in wildlife and its conservation. It is also hoped that primate tourism may generate income for local people and for Indonesia in general.

The book has benefited from the efforts of my former student, Priscillia Rindang and Indartono Sosro Wijoyo, who helped to collect and compile recent information concerning the ecology and behavior of Indonesian primates. Prof. Chris Margules from James Cook University of Australia and the University of Indonesia reviewed and edited the text. Dr. Russ Mittermeier and Dr. Anthony Rylands inspired me to write this book. Noel Rowe and many of my colleagues supplied primate photographs and Stephen Nash drew the pictures of each primate. Dr. Myron Shekkele, Prof. Sharon Gursky, Prof (riset) Ibnu Maryanto, and Late Prof Colin Groves supplied information and knowledge from their own experiences with Indonesian primates. I am grateful to all of these people.

ACKNOWLEDGMENT

This book will not be published without supports from many of my colleagues in Indonesia and other countries. I would like to express my special thanks of gratitude to my mentors, Prof Dr. Soekarja Somadikarta and Prof Dr. Indrawati Gandjar who helped and encouraged me a lot since I was a junior staff at Biology Dept, FMIPA, the Universitas Indonesia. I would also like to thank to my colleagues at Dept of Biology FMIPA, the Universitas Indonesia who have developed a better atmosphere at the workplace, especially to my research group on Wildlife and Sustainable Landscape, and the Vice Rector of the University of Indonesia, Prof Dr. Abdul Haris. I would like also to thank my collaborators in many NGOs such as UID Foundation Board of Trustee members (Cherie Nursalim, Sri Indrastuti Hadiputranto, Prof. Mari Pangestu, Aristides Katoppo, Marzuki Usman, I Gede Ardika, Laksamana Muda (Pur) Rosihan Arsyad, Prof Martani Husein, Ambasador Sudrajat), Belantara Foundation (Elim Sritaba, Agus Purnomo, Suhendra, Dr. Tahrir Fathoni, Dr. Ismayadi Syamsudin, Prof. Purwanto, Dr. Dolly Priatna, Dr. Sri Mariati). Thank you so much to my colleague primatologists, Dr. Russel Miitermeier, Dr. Anthony Rylands, Stephan Nash, Late Prof Don J. Melnick, Prof Jeffrey Froehlich, Prof Sharon Gursky, Dr. Myron Shekelle, Dr. Ramesh Boonratana, Dr. Lynn Clayton, Dr. Stefan Merker, Prof. Badrul Munir. I am also benefited from Indonesia's Primatologists and Conservation specialists who helped me getting more information for the book, Ir. Wiratno, M.Sc, Dr. Suci Utami, Dr. Noviar Andayani, Dr. Anton Ario, Dr. Sofyan Iskandar,

Dr. Yaya Rayadin, Dr. Tatang Mitra Setia, Dr. Jito Sugardjito, Dr. Barita Manullang, Dr. Nurul Winarni, Dr. Sri Mariati, Dr. Jamartin Sihite, Tantyo Bangun, Arif Setiawan, Diah Asri, Maya Dewi and many others that I cannot mention one by one. Thank you so much also to my assistants, Indartono Sosro Wijoyo and Priscillia Rindang, who help me finding the source of pictures and literatures.

Pictures and Drawings Credit

Drawing of all species were provided by Stephan Nash. Pictures were provided by various sources, personal documentation, and published materials. We would like to acknowledge those who provided drawings and pictures:

Anargha Setiadi (Research Center for Climate Change - University of Indonesia), Anton Ario, Ardika Dani Irawan, Arif Rudianto, Asri Ali, Badrul Munir (Universiti Kebangsaan Malaysia), Chairunisa Adha Putra, Denny Setiawan (International Animal Rescue), Ehler Smith, Ferdi Rangkuti, FX Ngindang, Gusmardi Indra, Heribertus Suciadi (International Animal Rescue), Ibnu Maryanto (LIPI), Indra Hana, Indra Yustian (Universitas Sriwijaya), James Kumolontang, Jarot Arisona, Kayan Mentarang National Park, Kristana Makur, Kuswanda, Lynn Clayton, Michel Gunther, Misdi, M. Indrawan (Research Center for Climate Change - University of Indonesia), M. Khotiem, M. Rizki Gumelar, Myron Shekelle, Nanang K. Hadi, Noel Rowe (Primate Conservation Inc.), Nur Rohman, Nurul L. Winarni (Research Center for Climate Change - University of Indonesia), Putu Sutarya, Randi Syafutra, Reki Kardiman and Risanti (International Animal Rescue), Sharon Gursky (Texas A&M University), Sofian Iskandar (Ministry of Environment and Forestry), Sri Mariati (Belantara), Stefan Merker, Sri Suci Utami Atmoko (Universitas Nasional), Suprayitno, Susan Cheyne (Borneo Nature Foundation), Tatang Mitra Setia (Universitas Nasional), Usman Wildlife Conservation Society -Indonesia Program.

Contents

1. An Overview of Indonesia's Primates

Indonesia has one of the richest primate faunas in the world. Sixty one species of the world's approximately 479 species (634 primate taxa including sub-species) occur in Indonesia. It ranks third in terms of number of species after Brazil (116 species) and Madagascar (98 species). They come from five families and 11 genera. Thirty eight are Indonesian endemics (Mittermeier et al., 2013). The primates are distributed across the archipelago from North Kalimantan to the south coast of Java, and from the westernmost parts of Sumatra east to Bacan and East Timor. Evolutionarily, Indonesia contains primates of every type, from primitive 'living fossils' such as tarsiers through to the very advanced apes, both the lesser apes (gibbons), and great apes (orangutan), which are closely related to humans.

On the Mentawai islands all of the primate species are endemic. There are four species on the large island of Siberut, the Siberut langur, *Presbytis siberu*, Kloss' gibbon, *Hylobates klossii*, the Siberut pigtailed macaque, *Macaca siberu* and an unusual langur, the pig-tailed langur, *Simias concolor*. There are also four on the smaller islands of the archipelago, North and South Pagai and Sipora. Two of them, the pigtailed langur, *P. concolor* and Kloss' gibbon, H. klossii also occur in Siberut but the other two, the Mentawai pig-tailed macaque, *Macaca pagensis* and the Mentawai langur, *Presbytis potenziana* are only found on these smaller islands.

The island of Sulawesi has a level of importance in primate endemicity that is similar in kind, though not in scale, to that of Madagascar. Sulawesi and its adjacent islands are home to eight macaque species and at least nine species of tarsier. These 15 species comprise more than 5% of all primate diversity and astonishingly 100% are endemic to Sulawesi. Of the eight macaque species, four are classified as threatened; the black macaque, *Macaca nigra,* the moor

J. Supriatna, *Field Guide to the Primates of Indonesia*,
https://doi.org/10.1007/978-3-030-83206-3_1

macaque, *Macaca maura,* heck's macaque, *Macaca hecki,* and the booted macaque, *Macaca ochreata.* The tarsiers have a different story to tell. Field research suggests that there may be many undescribed species. The vertical distribution of tarsiers over short distances has been observed in Lore Lindu National Park, where populations of Dian's tarsier, *Tarsius dentatus,* and the Sulawesi mountain tarsier, *Tarsius pumilus,* probably border or overlap vertically in the forest. But Jatna's tarsier first described from the Gorontalo forest in 2016 by Shekelle et al., (2016) indicates the possibilities of more new species to come. Using playback of calls and genetic analysis of specimens, they were able to identify a new species and similar methods may well turn up more.

Sumatra and Kalimantan share many primate species. The ancestors of all primates found on these islands originally came from mainland South-East Asia and have since differentiated locally. Sumatra has several endemic primates, such as the Sumatran grizzled langur, *Presbytis thomasi,* the banded langur, *Presbytis femoralis, and* the Sumatran white-handed gibbon, *Hylobates lar.* In Kalimantan, there are endemic primates such as the maroon gibbon, *Presbytis rubicundus,* the white-fronted gibbon, *P. frontata,* hose's langur, *P. hosei,* Miller's langur, *P. canicrus,* the Bornean gibbon, *Hylobates muelleri.* The Bornean white-bearded gibbon, *H. albibabris,* the East Bornean grey gibbon, *H. funereus* and Abbott's grey gibbon, *H. abotti.*

Java represents the southern limit of primate distribution in Asia. Some of the primates living in Sumatra and Kalimantan have become extinct in Java, for example, pigtailed macaques, orangutans and tarsiers. In Java there are four species of endemic primates, which are all classified as endangered; the Javan gibbon, *Hylobates moloch,* the Javan langur, *Presbytis comata,* the Javan leaf monkey, *Trachypithecus margiratus* and the Javan coucang, *Nycticebus javanicus.* Two other primate species found in Java are the ebony leaf-monkey, *Trachypithecus auratus,* and the long-tailed macaque *Macaca fascicularis.*

The high endemicity of Indonesia's primates is mirrored by their diversity in morphology, ecology, and behavior. The tarsiers have sometimes been called 'living fossils'. The proportions of the limbs, which indicate their tree-hoping gait, are very similar to those of earlier primates of the Eocene period. They have a long scratching claw on their third toe as well as a less pronounced one on their second toe. Enormous eyes and large ears indicate their nocturnal habits. The tail is long with a fine hair at the tip, which is used for support. Tarsiers can turn their heads through 180 degrees and look directly backward. They also have the ability to turn their bodies through 180 degrees in mid-air when leaping.

Tarsiers are one of the smallest monkeys in the world with adult male body weight at only 71-120 grams. Spectral tarsiers, *Tarsius spectrumgurskyae,* which are found in northern Sulawesi, live in small territorial family groups. Families have stable monogamy, and family size ranges from 3 to 7 individuals. Two exceptions to the rule of monogamy are found in the Philippine tarsier, *Tarsius syrichta,* which only occurs in the Philippines and the Bornean tarsier, *Tarsius borneanus,* which are polygamous. The Sulawesi mountain tarsier, *Tarsius pumilus,* which is smaller, darker and has longer nails than the spectral tarsier lives in moist forest more than 2.000 above sea level.

There are six species of slow loris in Indonesia; the Sunda slow loris, *Nycticebus coucang* in Sumatra, the Bangka slow loris, *N. bancanus* in the Bangka Belitung isalnds, the Bornean slow loris, *Nycticebus menagensis,* the Kalimantan slow loris, *N. borneanus,* and the Kayan River slow loris, *N. kayan* in Kalimantan and the Javanese slow loris, *Nycticebus javanicus,* found only in Java. Other species can be found throughout South-east Asia. Slow lorises are small, nocturnal, slow moving primates similar in behavior to the sloths of South America. Previously, lorises were presumed to be solitary primates, but recent studies using radio telemetry have revealed that they are actually highly social in their behavior. Lorises have been observed to sleep in social groups of up to seven individuals and to have overlapping home ranges.

Leaf monkeys (*Presbytis* and *Trachpithecus*) are remarkable for their color variation, of the fur, crown, tail, head and limbs ranging from black, white, brown, and red to mixtures of two colors. In Sumatra, there four species of leaf monkeys from the white of the mitred langur, *Presbytis mitrata*, to the orange colored black-crested Sumatran langur, *Presbytis melalophos*, the black Sumatran langur, *P. sumatrana*, and black and white langur, *P. bicolor*. Other leaf monkeys such as the banded langur, *Presbytis femoralis*, the Sumatran grizzled langur, *Presbytis thomasi*, and the pale-thighed langur, *Prebytis siamensis,* have markings that resemble eye glasses. One species found in Kalimantan and Sumatra, the silvered langur, *Trachypithecus cristatus*, has grey to black coloration.

Five species of leaf monkey are endemic to Indonesian Borneo. They are the red leaf monkey, *Presbytis rubicundus*, Hose's leaf monkey, *P. hosei*, Miller's langur, *P. canicrus,* the cross-marked langur, *P. chrysomelas*, and the white fronted leaf monkey, *P. frontata*. In Java, there are two endemic leaf monkeys, the Javan langur, *Presbytis comata*, and the west Javan langur, *Trachypithecus masrgiratus*. The ebony leaf monkey, *T. auratus* also occurs on Central and East Java and east to Bali and Lombok Islands. Leaf monkeys usually live in small groups of 10-30 individuals, with only monogamous pairs within the group. Although most leaf monkeys live in large groups with multiple males, some species are found in small groups of fewer than 15 individuals with only one adult male. An example is the Javan langur, *Presbytis comata*.

Macaques are the most widespread genus of monkeys in the world with a total of 20 species found from the African deserts to the snowy mountains of Japan. There are 11 species in Indonesia, occurring in Sumatra, Kalimantan, Java, Sulawesi, and the islands of the east all the way to Timor. The long tailed macaque, *Macaca fascicularis*, is distributed widely through Thailand, Malaysia, the Philippines, and Indonesia, while the Sunda pig-tailed macaques (*Macaca nemestrina*) occur on the Malay peninsula, Kalimantan and Sumatra. The Siberut pig-tailed macaque (*Macaca siberu*) and Mentawai pig-tailed macaque (*Macaca pagensis*), endemic to the Mentawai islands, share many characteristics with other pig-tailed macaques but are smaller and genetically different. Sulawesi macaques have distinctive forms in comparison to the rest of the macaque genus. In the past, the Sulawesi macaques have been assigned to eight species that vary from the brown Moor macaque, *Macaca maura,* in the south to the more varied faces, crests, and colors of the Tonkean macaque, *M. tonkeana*, with a white cheek, to Heck's macaque, *M. hecki* and the Gorontalo macaque, *M. nigricens* with prolonged snouts, to the black color and high crest of the black macaque, *Macaca nigra,* in North Sulawesi. In southeast Sulawesi, the booted macaque, *M. ochreata*, which has very distinctive *ischial callosities*, on the animals' buttocks, can be found.

The gibbons and siamangs are brachiators, which means they use their arms to move from tree branch to tree branch and they dangle by their arms and feed at the ends of swaying branches. They live in monogamous family groups and sing in sex specific choruses; males chorusing before sunrise and females later in the morning. The Siamang, the largest gibbon, occurs in Sumatra and on the Malay peninsula. Both sexes are almost black in color. They chorus in the morning and can be heard for kilometers in the jungle. Lar gibbons are found in North Sumatra and agile gibbons in Central Sumatra. The Bornean white-bearded gibbon, *Hylobates albibabris*, is found in West and South of Central Kalimantan. It is distinguished from Muller's gibbon, *Hylobates mulleri*, by the whiskers. Muller's gibbon has been split recently into three species, *H. mulleri, H. funeures* and

H. abotti. In the upper reaches of two rivers, Barito and Mahakam, a hybrid form of morphology and songs was also found.

The orangutan is the only great ape in Indonesia. Sumatran and Bornean orangutans were believed to be two sub-species of *Pongo pygmaeus*, but recent studies have shown that the Sumatran orangutan (*Pongo abelii*) is distinct from its Bornean relative (*Pongo pygmaeus*). There are three sub-species of the Bornean orangutan, all found on the island of Borneo: *Pongo pygmaeus pygmaeus* in Sarawak and Northwest Kalimantan, *Pongo pygmaeus wurmbii* in Southwest and Central Kalimantan, and *Pongo pygmaeus morio* in East Kalimantan and Sabah. The newly discovered orangutan in Tapanuli, North Sumatra, *Pongo tapanuliensis*, is very distinctive with curly fur in comparison to other orangutans in Sumatra and Kalimantan.

The orangutans have brown fur and move slowly either through the branches or knuckle-walking on the ground. Their locomotion is a modified form of arm-swinging; they suspend themselves by their arms or their legs. The male, weighing up to 75 kg, is much larger than the female and twice as heavy. Adult males have large cheek pads and their arms are much longer and stronger than their legs.

The orangutan is a rare arboreal ape and is confined to primary tropical rain forest throughout its range, which stretches from swampy forest to hill and mountainous forest at between 1,000 to 2,000 m above sea level. Its diet is predominantly fruit, although it also eats leaves, bark, and insects. Recent field observations suggest that they may also eat small animals and have been seen to feed on the carcass a dead gibbon. There has been intensive research on the behavior and ecology of orangutans in both Central and East Kalimantan and north Sumatra

2. Primate History, Primate-Watching and Primate-Lifelisting

Most people are familiar with primates, or monkeys and apes. Primates are mammals that separated from the primitive mammalian stock some 65 million years ago. Like other mammals they have very advance reproductive systems. The young are protected within the mother's body before birth and are nourished by the secretions of special glands, the mammary glands, after birth. Mammals are warm blooded, with fur-clad bodies to insulate them from the cold.

As a group, primates have complex and advanced brains compared with other mammals. They also have a well-developed visual sense, and, apart from *Homo sapiens*, a body that is specially adapted for living in trees. They can also function successfully in other habitats such as deserts, snowy mountains and grasslands. Most live in the tropical forests of Africa, Asia and South America. But there are some that live in cold mountains, such as the Japanese macaque in Fukushima and the Golden leaf monkey in China. Others, such as the Baboon, are found in savannahs, while the Gibraltar macaques live in a karst environment in southern Europe.

Primate watching is not as widespread as bird watching, which in the USA alone involves as many as 40 million people generating significant eco-tourism business. If primate watching was to increase in popularity, it too might create significant eco-tourism business opportunities. Dr. Russell A. Mittermeier first proposed the idea of primate watching, or what he called Primate life-listing. In his book "Lemurs of Madagascar", he introduced the idea of Lemur-watching and Lemur Life-listing as a new hobby that aimed to copy the success of bird-watching. Birds can be seen everywhere including cities and gardens as well as more natural habitats. However, primates can only be seen in the wild if you go to where the primate habitat is, mostly in the tropical forests of Asia, Africa and South America so a lot more effort is involved in primate watching. No doubt, fewer people will

J. Supriatna, *Field Guide to the Primates of Indonesia*,
https://doi.org/10.1007/978-3-030-83206-3_2

become primate watchers, but the rewards are commensurate with the effort. When you make the effort and go 'primate watching' in Indonesia, make a list of what you see and note where you saw it. Once you have this checklist, please email it to me (jatna.supriatna@ gmail.com or jsupriatna@sci.ui.ac.id) and I can upload it onto the primates of Indonesia website (www.primataIndonesia). This will increase our knowledge and understanding of Indonesian primates and also provide information that can be used to protect and conserve the primates in the wild.

Why are primates interesting to humans?

Primates are man's closest living relatives. They resemble each other much more closely than they resemble other mammals. This similarity is the result of a common inheritance. They look very human and often seem to behave in human ways. They play, investigate, manipulate new objects, learn fairly quickly and communicate with each other. Some of them sometimes use tools to obtain food and even occasionally make those tools. Primates form complex social groups and develop behavioral patterns which are often similar to the structure of human societies.

Undoubtedly many people feel this similarity to be disturbing. Primates are at the same time too much like people for the relationship to be ignored and too different for it to be freely acknowledged. Some find the appearance weird, or even disgusting, and subconsciously reject their close relationship. Others tend to attribute human reactions and motives to primates which can be misleading, as they are not conditioned by human ethics and prejudices. Each species has its own patterns of behavior.

Most people are only able to see primates in captivity, in zoos or circuses where they live abnormal lives in unnatural surroundings. To see them at their best and to appreciate their natural behavior and beauty, it is important to observe them in their natural habitats.

3. Using This Field Guide

This primate field guide can be used to refer to information on each species, or it can be used to find which species exist on each island, as shown at the back of the book. The list of primates in Indonesia is provided with local, English and scientific names. Once you have identified the name, then you can go to the description of the genus and species. I have also given the conservation status of each species except for the most recently described, whose status is not yet known. To each species description has also been added information on its natural history, behavior, ecology, and where to see it in parks and/ or forested areas outside parks. Primate drawings were provided by Stephen Nash and photographs were donated by many friends from Indonesia and abroad.

© The Author(s), under exclusive license to Springer Nature Switzerland AG 2022
J. Supriatna, *Field Guide to the Primates of Indonesia*,
https://doi.org/10.1007/978-3-030-83206-3_3

Tabel 1. Classification of Indonesian Primates

A. Family Lorisidae (Gray, 1821)
Nycticebus (Geoffroy, 1821)
1. *N. coucang* (Boddaert, 1785) Sunda Slow Loris
2. *N. javanicus* (Geoffroy, 1812) Javan Slow Loris
3. *N. menagensis* (Trouessart, 1987) Bornean Slow Loris
4. *N. bancanus* (Lyon, 1906) Bangka Slow Loris
5. *N. borneanus* (Lyon, 1906) Bornean Slow Loris
6. *N. kayan* (Munds et al., 2013) Kayan River Slow Loris

B. Family Tarsiidae (Gray, 1825)
Tarsius (Erxleben, 1777)
7. *T. tarsier* (Erxleben, 1777), Selayar Tarsier
8. *T. fuscus* (Fischer, 1804), Makassar Tarsier
9. *T. dentatus* (Miller& Hollister, 1921), Dian's Tarsier
10. *T. pelengensis* (Sody, 1949), Peleng Tarsier
11. *T. sangirensis* (Meyer, 1897), Great Sangihe Tarsier
12. *T. tumpara* (Shekelle et al, 2008), Siau Island Tarsier
13. *T. pumilus* (Miller & Hollister, 1921), Sulawesi Mountain Tarsier
14. *T. lariang* (Merker & Groves, 2006), Lariang Tarsier
15. *T. wallacei* (Merker et al., 2010) Wallace's Tarsier
16. *T. supriatnai* (Shekelle et al, 2017), Jatna's tarsier.
17. *T. spectrumgurskyae* (Shekelle et al, 2017), Gursky's tarsier

Cephalopachus (Swainson, 1835)
18. *C. bancanus* (Horsfield, 1821) Western Tarsier

C. b. bancanus (Horsfield, 1821) Horsfield's Tarsier
C. b. natunensis (Chasen, 1940) Natuna Islands Tarsier
C. b. saltator (Elliot, 1910) Belitung Tarsier
C. b. borneanus (Elliot, 1910) Bornean Tarsier

C. Family Cercopithecidae
Macaca (Lacépède, 1799)
19. *M. nemestrina* (Linnaeus, 1766) Sunda Pig-tailed Macaques
20. *M. siberu* (Fuentes & Olson, 1995) Siberut Macaque
21. *M. pagensis* (Miller, 1903) Pagai Macaque
22. *M. nigra* (Desmarest, 1822) Crested Macaque
23. *M. nigrescens* (Temminck, 1849) Gorontalo Macaque
24. *M. tonkeana* (Meyer, 1899) Tonkean Macaque
25. *M. ochreata* (Ogilby, 1841) Booted Macaque

M. o. ochreata (Ogilby, 1841) Booted Macaque
M. o. brunnescens (Matschie, 1901)
26. *M. hecki* (Matschie, 1901) Heck's Macaque
27. *M. maura* (Schinz, 1825) Moor Macaque
28. *M. fascicularis* (Raffles, 1821) Long-tailed Macaque

M. f. fuscus (Miller, 1903) Simeuleu Long-tailed Macaque
M. f. karimondjawae (Sody, 1949) Kemujan Long-tailed Macaque
M. f. lasiae (Lyon 1916) Lasia Long-tailed Macaque
M. f. tua (Kellog, 1944) Maratua Long-tailed Macaque
Presbytis (Eschsholtz, 1821)
29. *P. thomasi* (Collet, 1893) Thomas's Langur
30. *P. melalophos* (Raffles, 1821) Black-crested Sumatran Langur
31. *P. sumatrana* (Müller & Schlegel, 1841) Black Sumatran Langur
32. *P. bicolor* (Aimi & Bakar, 1992) Black-and-white Langur
33. *P. mitrata* (Eshscholtz, 1821) Mitred Langur
34. *P. comata* (Desmarest, 1822) Javan Langur

P. c. comata (Desmarest, 1822) Javan Grizzled Langur
P. c. fredericae (Sody, 1930) Javan Fuscous Langur
35. *P. potenziani* (Bonaparte, 1856) Pagai Langur
36. *P. siberu* (Chasen & Kloss, 1928) Siberut Langur

37. *P. femoralis* (Martin, 1838) Banded Langur
P. f. percura (Lyon,1908) eas Sumatran Banded Langur
38. *P. siamensis* (Müller & Schlegel, 1841) Pale-thighed Langur

P. s. cana (Miller, 1906) Riau Pale-thighed Langur
P. s. paenulata (Chasen, 1940) Mantled Pale-thighed Langur
P. s. rhionis (Miller, 1903) Bintan Pale-thighed Langur
39. *P. natunae* (Thomas & Hartert, 1894) Natuna Islands Langur
40. *P. chrysomelas* (Müller, 1838) Cross-marked Langur

P. c. chrysomelas (Müller, 1838) Bornean Cross-marked Langur
41. *P. rubicunda* (Müller, 1838) Maroon Langur

P. r. rubicunda (Müller, 1838) Maroon Red Langur
P. r. carimatae (Miller, 1906) Red-naped Langur
P. r. ignita (Dollman, 1909) Orange-naped Red Langur
P. r. rubida (Lyon, 1911) Southwest Kalimantan Red Langur
42. *P. hosei* (Thomas, 1889) Hose's Langur
43. *P. canicrus* (Miller, 1934) Miller's Langur
44. *P. frontata* (Müller, 1838) White-fronted Langur

Trachypithecus (Horsfield, 1823)
45. *T. auratus* (Geoffroy, 1821) East Javan Langur
46. *T. mauritius* (Griffith, 1821) West Javan Langur
47. *T. cristatus* (Raffles, 1821) Silvered Langur

T. c. cristatus Raffles 1821) Common Silvered Langur
T. c. vigilans (Miller, 1913) Natuna Islands Silvered Langur
Nasalis (Geoffroy, 1812)
48. *N. larvatus* (van Wurmb, 1787) Proboscis Monkey

Simias (Miller, 1903)

49. *S. concolor* (Miller, 1903) Pig-tailed Langur

S. c.concolor (Miller, 1903) Pagai Pig-tailed Langur
S. c. siberu (Chasen & Kloss, 1928) Siberut Pig-tailed Langur

D. Family Hylobatidae
Hylobates (Illiger, 1811)
50. *H. lar* (Linnaeus, 1771) Lar Gibbon
H. l. vestitus (Miller, 1942) Sumatran Lar Gibbon
51. *H. agilis* (Cuvier, 1821) Agile Gibbon
52. *H. albibarbis* (Lyon 1911) Bornean White-bearded Gibbon
53. *H. muelleri* (Martin, 1841) Müller's Gibbon
54. *H. abbotti* (Kloss, 1929) Abbott's Grey Gibbon
55. *H. funereus* (I. Geoffroy Saint-Hilaire, 1850) East Bornean Grey Gibbon
56. *H. klossii* (Miller,1903) Kloss's Gibbon
57. *H. moloch* (Audebert, 1798), Javan Gibbon

Symphalangus (Gloger, 1841)
58. *S. syndactylus* (Raffles, 1821) Siamang

E. Family Pongidae
Pongo (Lacépède, 1799)
59. *P. abelii* (Lesson, 1827) Sumatran Orangutan
60. *P. pygmaeus* (Linnaeus, 1760) Bornean Orangutan

P. p. pygmaeus (Linnaeus, 1760) North-west Bornean Orangutan
P. p. morio (Owen, 1837) North-east Bornean Orangutan
P. p. wurmbii (Tiedemann, 1808) South-west Bornean Orangutan
61. *P. tapanuliensis* (Nater et al., 2017), Tapanuli orangutan, Sumatra.

A. FAMILY LORISIDAE (Gray, 1821)

This family includes two genera, but only the genus *Nycticebus* is found in Indonesia. The other genus, Loris, occurs in India and Sri Lanka. Previously, Groves (2001) classified the genus *Nycticebus* into five species, *N. bengalensis*, *N. pygmaeus* and three species from Indonesia *N. coucang*, *N. menagensis* and *N. javanicus*. A recent review of museum specimens and photographs in 2013 elevated two new species (*N. bancanus* and *N. borneanus*) from former subspecies of *N. menagensis* and proposed an additional new species, *N. kayan*. However, confirmation of these findings awaits a comprehensive genetic study of Bornean *Nycticebus* populations (Munds et al., 2013).

Indonesian slow lorises range in body weight from the largest, *N. javanicus* at 0.685 kg to the smallest, *N. menagensis* at 0.265 kg. They have a wet nose that helps improve olfactory function. They also have a toilet claw on their second digit and a tooth comb for grooming. The tooth comb is a dental structure comprising four procumbent lower incisors and two lower canines. As well as their use for grooming purposes, tooth combs are also used in the gouging behavior needed for feeding on gum. All lorises are arboreal. But a study in Talun plantation in Garut, West Java, revealed that lorises are also capable of moving on the ground when continuous tree cover is not available (Putri, 2014). To move from one tree to the other, lorises climb slowly, stretching their body as they don't have the capability to jump or leap. In the field, you can easily recognize lorises by their nocturnal nature and big eyes that reflect red to orange eye shine when spotlighted. They can also be found in the daylight

J. Supriatna, *Field Guide to the Primates of Indonesia*,
https://doi.org/10.1007/978-3-030-83206-3_4

while sleeping, with their bodies rolled into a "sleeping ball" posture trying to camouflage themselves in the forest.

In general, their body hair color varies from dark brown to reddish brown with a distinctive facial coloration pattern. Out of all the lorises in Indonesia, *N. javanicus* has the most distinct facial pattern with a bold white diamond shaped by dark brown stripes that run over its eyes and ears (Fig 3). In Bornean populations, these facial patterns are believed to be important for distinguishing species and also for identifying potential mates and conspecifics.

Lorises are the only venomous primates in the world. All species have a brachial gland under their arm, which excretes oil that can be activated to become poisonous when combined with saliva. Slow lorises spread this toxin across their bodies and those of their offspring while grooming with their toothcombs. Loris venom may have originated as an anti-parasitic defense. It may also serve to avoid olfactory-oriented predators such as cats, sun bears and civets (Nekaris et al., 2013). When threatened by a predator, they can bite with this toxin, which can cause death in small mammals and anaphylactic shock and even death in humans (Madani & Nekaris, 2014). However, their primary defence mechanism is crypsis, whereby they avoid observation or detection by hiding.

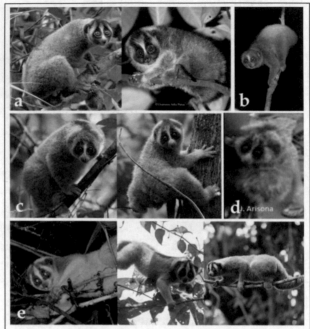

Figure 1. Indonesian slow lorises: a. *Nycticebus coucang*
(Denny Setiawan - IAR) & Infant (Chairunisa Adha Putra);
b. *Nycticebus bancanus* (Randi Syafutra), c. *Nycticebus
menagensis* (Heribertus Suciadi - IAR); d. *Nycticebus
borneanus* (Jarot Arisona); e. *Nycticebus javanicus* (Anton
Ario, Denny Setiawan – IAR, & Risanti - IAR).

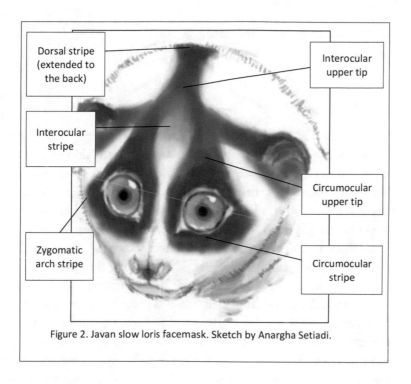

Figure 2. Javan slow loris facemask. Sketch by Anargha Setiadi.

1. Sunda Slow Loris

Nycticebus coucang

Other common names: Malu-malu, Kukang
Sumatra, Bukang, Kalamasan

Identification

The head and body length of the Sunda slow loris
is about 28.2 - 31.7 cm with a short vestigial tail hidden beneath
the fur. Adults range in weight from 0.599 - 0.685 kg (Nekaris et al.,
2008). They have short, thick, wooly fur. In general, coat coloration
is reddish brown with a lighter brown underside. They have a grayer
chest color compared to the belly. There are dark circumocular stripes
and a lightly whitish stripe that extends to the forehead. There is a dark
dorsal stripe from the top of the head back along the spine (Groves,
2001; Gron, 2009). The color and shape of the circumocular fork and
stripe prove to be useful in identifying this species (Mittermeier et al.,
2013). Compared to the larger Bengal slow loris, the sunda slow loris
has a more distinct brown coat and darker circumocular stripes, a
more distinct white stripe, and a darker dorsal stripe. It has less white
facial coloring compared to the smaller pygmy slow loris and a less
bold circumocular stripe compared to the Javan slow loris. The ears
are short.

Geographic Range

Sunda slow lorises are found throughout much of Southeast Asia. In Indonesia they are found throughout Sumatra, including Batam island, in the Riau Archipelago, and Bunguran in the North Natuna Islands. They are also found in southern peninsular Thailand, peninsular Malaysia and the island of Pulau Tioman (Groves, 2001; Roos et al., 2014).

Behavior and Ecology

Sunda slow lorises typically are active during the night and occur in primary or secondary forests and bamboo groves, which have a continuous dense canopy. They walk hand over hand along branches, bringing the hind foot forward to meet the hand. They commonly climb slowly and deliberately although they can strike with amazing speed, but they don't have the ability to jump (Ishida, 1992). They grip branches with both hind feet, stand erect and throw their bodies forward. Their slow movement is presumed to be related to their slow metabolic rate due to the energy cost of detoxifying certain secondary plant compounds or toxins in insects of their diet. Even though they have a slow metabolic rate, this species has a high-energy diet (Wiens et al., 2006). The diet is omnivorous and includes gum, sap, fruit, and floral nectar (Wiens, 2002). They use their lower anterior teeth to scoop up gum that has already been exuded and solidified (Streicher, 2004; Wiens et al., 2006).

Sunda slow lorises occur at very low densities. Studies reveal that lorises have extremely large home ranges relative to their body size (Wiens, 2002). Adults have over-lapping home ranges and form groups with friendly interaction among members. They have also occasionally been observed sleeping in groups (Wiens, 2002). These groups consist of one male, one female and up to three younger offspring. When they come into contact with conspecifics with other home ranges there is usually no reaction as home ranges are not defended. They also consume large molluscs, insects, lizards, birds, and small mammals. This species home range is around 10-25 ha.

Sexual maturity is reached between the ages of 18 and 24 months in females, and can be reached by 17 months in males. Females have an estrous cycle of 37 - 54 days and are reproductive for 5 - 6 days. There is usually a single young, occasionally twins, and births apparently occur in the open. Soon after birth, infants are left alone and will remain in the exact site where they were born. This is called "infant parking". Later when infants become mobile (in a few months), they are able to move around on their own (Napier and Napier, 1967; Wiens, 2002).

Conservation Status

The Sunda slow loris is listed as Vulnerable on the IUCN Red list and in Appendix I of CITES. It means that this species status is vulnerable with population size reduced to less than 30% of the original and still declining due to forest degradation and hunting. They are protected under Indonesian law. However, this does not stop hunting of this species, mainly for the pet trade. They are known to be sold throughout Southeast Asia. The Sumatran populations are particularly impacted by the pet trade. Before being sold in the market, their teeth are often pulled, which then can cause infection and sometimes lead to death. Moreover, the lack of teeth in individuals recovered from the pet trade makes reintroduction efforts to the wild difficult.

Where to See It

Sunda slow lorises can be seen in most forests and several protected areas throughout their range. They are found on Sumatra Island (including Batam and Galang in the Riau Archipelago, Tebing tinggi Island and Great Natuna in the Natuna Islands) (Groves, 2001). Studies indicate that Sunda slow lorises occupy the forest edge of Bukit Barisan National Park. Normally they occupy primary or secondary forests and bamboo groves. It may be difficult to observe this species due to its cryptic nature. It has a bright red to orange eye shine when spotlighted at night. Nekaris (1997) found that using a red light torch is less disturbing for the lorises than a bright white light. Walking slowly while sweeping the canopy with a red-light torch is the most likely way to see this species in its natural habitat.

2. Javan Slow Loris

Nycticebus javanicus

Other common names: Kukang Jawa (Indonesian), Muka, Muka Geuni, Oces, Aeud (Sunda).

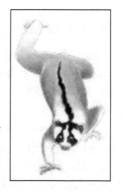

Identification

The Javan slow loris is the largest species of Indonesian loris with head and body length around 28 - 30.6 cm (Nekaris & Jaffe, 2007). Adult weight ranges from 0.5 - 0.7 kg (Nekaris et al., 2008). It has brown to reddish body coloration with white colored hair on the face and nuchal area (nape of the neck). A dark circumocular stripe extends from the cheeks to the forehead and runs back along the spine. Compared to other Indonesian slow lorises, Javan slow lorises have bolder facial markings. The contrast between the white hair of the face and the dark hair of the circumocular zone creates a distinct diamond shaped interocular area. This diamond shape is the defining characteristic of the Javan slow loris. They also have small ears with a tuft. Second incisors in their dentition are absent (Nekaris & Jaffe, 2007; Groves, 2001).

Geographic Range

Javan slow lorises are endemic to Java Island. Previously their distribution was thought to be restricted to West and Central Java, but a recent study by Lethinen (2013) confirmed their presence in Meru Betirin National Park in East Java.

Behavior and Ecology

Javan slow lorises feed on floral inflorescences, tree sap and gum, insects, fruit, lizards, and eggs (Wirdateti et al., 2004, 2011; Nekaris & Munds, 2010). At Cipaganti research station in West Java, they were observed mainly to fed on *Acacia deccurens* exudates (56%) and *Calliandra calothyrsus* nectar (32%) (Rode-Margono et al., 2014) while individuals reintroduced to Mt. Salak were seen mostly to feed on *Calliandra calothyrsus* nectar (90%) (Moore et al., 2011). On one occasion they have been seen to feed on *Diospyros kaki* fruit, which have a high tannin content that makes the unripe fruit taste bitter (Putri, 2014).

They occupy primary and secondary forest, as well as forest remnants and "Talun Plantations" practice in West and Central Java, which combines crops and tree plantations in agricultural areas (Winarti, 2008; Lethinen, 2013). Recent studies confirm the ability of this species to inhabit disturbed plantation areas and tolerate a high level of human disturbance, especially where the landscape includes significant amounts of bamboo (Putri, 2014). This species prefers a habitat with good tree connectivity in order to move but has also been observed crossing the ground when tree connections are unavailable (Winarti, 2008; Putri, 2014).

Javan Slow Lorises are active during the night. They spend most of their time foraging and feeding, and travelling (Rode-Margono et al., 2014). A study of Javan Slow Lorises in a plantation area revealed that they performed most of their activities between 1-5.9 m (53.51%) off the ground (Putri, 2014). In comparison, Arisona (2008) found that they were usually encountered between 5- 10 m above ground in Gunung Gede National Park. It is assumed that at the 1- 5.9 m height range, trees in plantation habitat are connected, which supports the travel behavior of this species, behavior that is also found in the slender loris (Nekaris et al., 2014, Putri, 2014). A high degree of tree connectivity is important for Javan slow lorises as they climb slowly from one tree to another. Like all other lorises, the Javan slow loris does not have the ability to leap. In plantations, the Javan slow loris has been seen to travel on the ground even using human artificial substrates such as water pipes and wire (Putri et al., 2013).

Conservation Status

The Javan slow loris was upgraded from Endangered to Critically Endangered by the IUCN redlist in 2014 due to the rapid decline of its population (IUCN list: Cr A2cd+4cd). This species is threatened by habitat loss and the illegal pet trade. Its distribution range in Java overlaps with the Sundaland Biodiversity Hotspot, which has experienced extensive deforestation with only an estimated 20% of its original forest remaining (Thorn et al., 2009). Furthermore, environmental niche modelling has shown that this species is among the most threatened slow lorises by habitat loss (Nekaris & Munds, 2010). Besides habitat loss, they also suffer from hunting for the illegal pet trade and traditional medicine (Nekaris et al., 2010). Their relatively slow locomotion, nocturnal nature, tendency to sleep on open branches, and freezing behavior when threatened by predators make them easy to catch.

Where to See It

The Javan Slow Loris can be seen in primary and secondary disturbed lowland to highland forest, bamboo forest, mangrove, plantations, especially Cocoa plantations, within their range from West to East Java (Supriatna & Wahyono, 2000). They are found in several protected areas such as Ujung Kulon National Park, Halimun Salak National Park, Gunung Gede-Pangrango National Park, Gunung Masigit-Kareumbi Game Reserve and many other places in Western Java. This slow loris could still be found even in Jakarta, in the Kemayoran forest and at the University of Indonesia campus a few years ago.

3. Bornean Slow Loris

Nycticebus menagensis

Other common names: Kukang Kalimantan (Indonesian), Kalamasan (Banjar)

Identification

The body weight of this species ranges between 0.265 – 0.610 kg but an animal rescued from the pet trade was found to weigh 0.8 kg. The average body length is 27.42 cm (Nekaris & Munds, 2010). Bornean slow lorises have a paler facemask compared to other Slow Lorises in Indonesia with a rounded diffuse-edge upper circumocular patch. This patch is variable and sometimes extends to below the eyes (zygomatic arch).

Geographic Range

This Loris can be found throughout North Kalimantan, Sabah and Sarawak, comprising the island of Borneo, and also in the Phillipines.

Behavior and Ecology

Similar to other lorises, this species consumes mainly insects, fruit, and tree gum. They travel and search for food using a relatively slow movement in the top of the tree canopy, usually between 10-15 m in height. The density is very low, about 0.21-0.38 animals/km2 (Nekaris and Jaffee, 2008). Surveys have found encounter rates of 0.02 individual/km (Nekaris and Munds, 2010). More research needs to be done to better understand the behavior and ecology of Bornean slow lorises.

Conservation Status

Nycticebus menagensis is listed as Vulnerable in both the IUCN Redlist and CITES Appendix I. IUCN listed it with criteria VU A2cd+3cd with a more than 30% reduction in population size over three generations.

Where to See It

N. menagensis can be found in several national parks such as Kutai NP in East Kalimantan. Kutai NP can be reached by flying to Balikpapan and then following one of the routes below:

- Balikpapan ──→ Samarinda (125 km), which takes up to 2 hours by car or a 15 minute flight.
  - o Samarinda (Lampake) ──→ Bontang, which takes 4 hours by car
  - o Bontang ──→ Sanggata (50 km). This route can be taken with a 30 minute flight or motor boat from Lok Tuan harbor through Tanjung Pandan, Teluk Kaba, Teluk Lombok and will take about 3 hours.
  - o Sangatta ──→ Mentoko using ketinting (Small boat)
- Alternatively, you can fly from Balikpapan to Bontang using plane of PT. Badak LNG and PT Pupuk Kaltim for 40 minutes.

4. Kalimantan Slow Loris

Nycticebus borneanus

Other common names: Kukang Kalimantan (Indonesian), Bengkang (Iban)

Identification

Formerly considered a subspecies or synonym of *N. menagensis,* it was promoted to full species status in 2013 when a study of museum specimens and photographs identified distinct facial markings, which helped to differentiate it as a separate species. It is distinguished by its dark, contrasting facial features, as well as the shape and width of the stripes of its facial markings. A molecular study analysis by Chen et al., (2003) of the genus *Nycticebus* showed *N. borneanus* to be genetically distinct from *N. coucang.*

N. borneanus is a medium-sized loris with an average head and body length of 26.0 cm (from four specimens). The facemask shows a dark contrasting with mostly round upper tip, but sometimes also with a diffuse-edged. The circumocular patch never extends below the zygomatic arch. The interocular stripe is variable in width and often found as a round or band shape, but never diffuse. They have hairy ears with a wide preauricular hair band (Nekaris and Munds, 2010, Munds et al., 2013).

Geographic Range

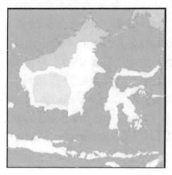

This species is found in West, South and Central Kalimantan, but excluding the extreme Southwest. *N. borneanus* is assumed to be sympatric with *N. bancanus* in West Kalimantan. It is speculated but still not known, if the Kapuas River plays the role of a barrier between these two species.

Behavior and Ecology

It was first described as *N. borneanus* from western Borneo by Marcus Ward Lyon in 1906. In 1953, all the slow lorises were lumped together into a single species, the Sunda slow loris (*N. coucang*). This was updated in 1971 when the pygmy slow loris (*N. pygmaeus*) was distinguished as a single species and four subspecies of *N. coucang*, including *N. coucang menagensis* were recognized. In 2006, *N. menagensis* was elevated to a single species when molecular analysis showed it to be genetically distinct from *N. coucang*. *N. borneanus* was given full species status in 2013 when it was differentiated from *N. menagensis* using museum specimens and photographs (Munds et al., 2013).

Conservation Status

While this new species has yet to be assessed by the IUCN, *N. menagensis* was listed as «Vulnerable» as of 2012 (Munds et al., 2013). Because one original species has been divided into four new species, each of the new species faces a higher risk of extinction. Accordingly, each of them is expected to be listed as "Vulnerable" at the least, with some of them likely to be assigned to a higher-risk category. Between 1987 and 2012, one-third of the area of Borneo's forests has been lost, making habitat loss one of the greatest threats to the survival of *N. borneanus*. The illegal wildlife trade is also a major factor, with loris parts commonly sold for use in traditional medicine and videos appearing on YouTube promoting the wildlife trade. However, all slow loris species are protected from commercial trade under Appendix I of CITES (Nekaris and Munds, 2010).

Where to See It

This species is found in tropical forest including primary, secondary, moist, mountain, evergreen, peat swamp, submontain evergreen, coastal lowland, riparian, dry coastal, gallery, and deciduous. Bornean slow lorises are not easy to detect due to their nocturnal habits and their slow movement. They usually hide behind the leaves which makes them harder to find. They occur near Kabupaten Sanggau in West Kalimantan and can be found in at least two national parks that already have primate research stations, Gunung Palung National Park (Rimbo Pati Research Station) and Tanjung Puting NP (Orangutan Camp Leakey Research Station). Gunung Palung NP can be reached by following the route below:

- Airport in Pontianak ⟶ Ketapang (1 hour and 15 mins flight). Then continue with rental car or boat for 6 – 10 hours. To reach the national park you will then have to hike for around 5 hours.

While Tanjung Puting NP can be reached through this route:
- Airport in Pangkalan Bun ———→ Kecamatan Kumai using rental car (1 hour) or rental motor bike (1.5 hours). To reach the national park, you can take the motor boat along the Kumai River and turn into the Sekonyer Kanan River (4 hours).

5. Kayan River Slow loris

Nycticebus kayan

Other common name: Kukang kayan (Indonesian)

Identification

The Kayan River Slow Loris has a distinctive facemask that distinguishes it from other Bornean lorises. The facemask has a distinct black and white contrast. The top dark circumocular is either rounded or pointed (not diffuse at the edges). The circumocular patch stretches below the zygomatic arch and often as far as the base of the lower jaw. The interocular stripe is almost always narrow and either rectangular or bulb shaped. The crown patch is variable but mainly diffuses along the edges. They also have hairy hair at the preauricular hair band. The average head and body length is 27.3 mm (from ten specimens) and the average body weight is 0.410 kg (from two specimens) (Munds et al., 2013).

Geographic Range

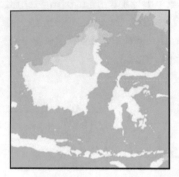

N. kayan is found in Central and Northern Borneo (Sarawak, Sabah, East Kalimantan and North Kalimantan). There is little accurate information on their precise geographic distribution. Further study is needed in order to ascertain the exact habitat type.

Behavior and Ecology

This species was recognized in 2013 based on a review of specimens and photos. Field studies are still lacking and there are no studies yet on its behavior and ecology (Munds et al., 2013). But it is expected to occur sympatric with *N. menagensis* in East Kalimantan and Sabah.

Conservation Status

Neither IUCN nor CITES have reviewed and listed the status of the species yet. But like other slow lorises, it is likely that habitat loss and the illegal wildlife trade affecting the survival of *N. kayan*.

Where to See It

Similar to other lorises, they are very hard to find due to their nocturnal habits, their slow movement and their propensity to hide when threatened. This species can be found in Kayan Mentarang National Park in North Kalimantan. To reach this national park you need to spend around 3 days on the Kayan River and then go in the direction of the Bahau River and you will arrive in the Nggeng estuary. There will be a welcome gate there and some accommodation available. From there you can hike for half an hour by three different routes to get to Laut Birai Estuary Alternatively, you can take a flight from Tarakan for 1 hour and continue with a smaller plane to Karayan, Punjungan or Apokayan. From Tarakan you can also rent a motor boat throughout the Mentarang River. With motor boat, you can reach Punjungan from Long Bio.

6. Bangka Slow Loris

Nycticebus bancanus

Other common names: Kukang bangka (Indonesian)

Identification

Bangka slow lorises have a distinctive crimson red dorsal pelage and medium body size (25.8 cm from six specimens). They have a facemask with diffuse-edged upper circumocular and lower circumocular stripes that never extend below the zygomatic arch. The interocular stripe is medium sized. They have hairy ears with narrow preauricular hair.

Geographic Range

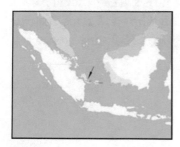

This species is found on the island of Bangka and also in West Kalimantan. The Bangka population is allopatric to all other slow lorises, but the Kalimantan population has some possible sympatry with *N. borneanus* in West Kalimantan.

Behavior and Ecology

N. bancanus was formerly recognized as a synonym or subspecies of *N. menagensis*. In 2013, Munds et al., (2013) recognized *N. bancanus* as a single species from museum specimens and photographs. The ecology and behavior of *N. bancanus* has not been studied yet. But just like other slow lorises, the diet includes fruits, supplemented with leaves, shots, saps, gums, flowers, seeds and insects. The Bangka slow lorises are nocturnal and arboreal.

Conservation Status

Neither IUCN nor CITES has so far reviewed and listed the status of this species. But like other slow lorises, it is likely that habitat loss and the illegal wildlife trade are affecting the survival of *N. bancanus*. Considering Bangka and Belitung are small islands on the east coast of Sumatra, the IUCN tentatively categorized them as Vulnerable, just like many other species of slow loris except the Javan species.

Where to See It

If you are visiting Bangka and Belitung Islands, there is a small forest where this species can be seen near the city of Muntok, Pangkal Pinang on Bangka or forest near Belitung city on Belitung island. You can also go to the Gunung Maras National Park. The national park is 16,806 ha. in size and is located in the Riau Silip and Kelapa Districts of Bangka and Bangka Barat Regencies, Bangka Belitung Province.

B. FAMILY TARSIIDAE (Gray, 1825)

Tarsiers are often called "living fossils". Tarsiers are one of the smallest primates in the world. Their body length ranges from 8.5 - 16.0 cm with tail length ranging from 13.5 - 27.5 cm. The body weight of mature males is approximately 0.075 - 0.165 kg. This early ancestor evolved in the Eocene era and is characterized by its unique foot form. Tarsiers generally can turn their heads through almost 360 and can see backwards without moving their body.

All species in this family are nocturnal. Their general characteristics are their big eyes and widely spaced ears compare to the size of their heads. They vary in color from dark red, to brown and grey depending on the species. The Tarsiers of Sulawesi have distinct features compared to other species of tarsier, which are white hair behind their ears, and grey body hair. The name Tarsier is connected to the Latin word for part of the foot (tarsus). The length of their feet is greater than the length of their hands. This is very much related to the way they move by jumping.

In the past century, there have been debates on the phylogenetic position of the family Tarsiidae. Tarsiers are almost invariably placed in the group with Simiiformes (= Anthropoidae). However, according to Murphy et al., (2001), they may be closer to Strepsirrhini. Strepsirrhines are defined by their wet nose or rhinarium. They also have a smaller brain than comparably sized simians, large olfactory lobes, a vomeronasal organ to detect pheromones and a bicornuate uterus with an epitheliochorial placenta

Previously, there was only one genus recognized in this family, the genus Tarsius (Groves, 2005). In 2010, Groves & Shekelle proposed

J. Supriatna, *Field Guide to the Primates of Indonesia*,
https://doi.org/10.1007/978-3-030-83206-3_5

a revised taxonomy and split the genus Tarsius into three genera, the Philippine tarsiers (genus *Carlito*), the western tarsiers (genus *Cephalopachus*), and the eastern tarsiers (genus *Tarsius*). The genus *Carlito* is only found in the islands of the Philippines while the genera *Cephalopachus* and *Tarsius* are found in Indonesia. The genus *Cephalopachus* is found in southern Sumatra and Borneo. This genus has only one species, *Cephalopachus bancanus*, but there are 4 subspecies, Horsfield's Tarsier (*C. b. bancanus*), the Natuna Islands Tarsier (*C. b. natunensis*), the Belitung Tarsier (*C. b. saltator*), and the Bornean Tarsier (*C. b. borneanus*). The other genus in Indonesia, *Tarsius* is spread across Sulawesi and its adjacent islands. There are nine species recognized in this genus (Shekelle et al., 2004).

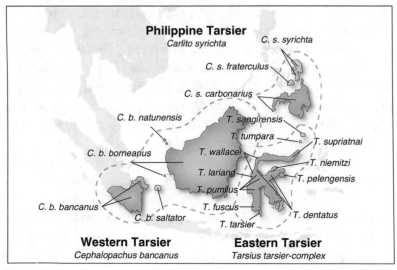

Figure 3. Distribution map of the genera and species of Tarsiidae. Source: Groves & Shekelle, 2010

The local names of tarsiers in Sulawesi and its surrounding islands are Tangkasi (Minahasa), Ngasi (Central Sulawesi), Singapuar, Tanda-bona, Pasoo (Wana), Podi (Tolaki), Wengu (Morenene), Tenggahe (Sangir), Tanda-bana (North Sulawesi). Lakasinding (West Peleng) and Siling (East Peleng). While in Sumatra the names are: Singapuar (Bengkulu); Krabuku (Lampung), Palele (Belitung), Mentilin Ingkir, Ingkit, Beruk-puar (Bangka): and Kalimantan: Ingkir, Linseng (Ngaju), Ingkat (Iban); Page (Tidung), Makikebuku (Karimata), Singaholeh (Kutai), Tempiling (Kalbar), Binatang hantu, Simpalili (Melayu).

When the small nocturnal tarsiers retire to their sleeping trees in the morning, they usually perform duet calls at or close to their sleeping sites. A sleeping tree is only occupied by one group usually consisting of one adult male, one adult female, and their offspring (MacKinnon & MacKinnon, 1980, Gursky, 1998). The morning duets have a territorial as well as a social component collecting the members of the groups (Nietsch & Kopp, 1998, Nietsch, 1999). Their call is a loud high pitched squealing.

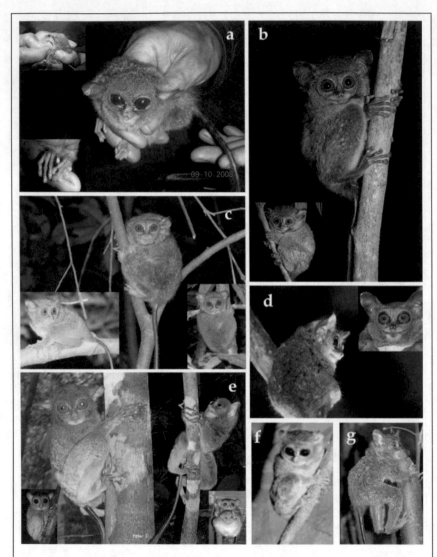

Figure 2. Indonesian Tarsier: a. *Tarsius pumilus* (Sharon Gursky); b. *Tarsius lariang* (Stefan Merker); c. *Tarsius spectrumgurskyae* (Sharon Gursky); d. *Tarsius wallacei* (Stefan Merker); e. *Cephalopachus bancanus* (Indra Yustian & Myron Shekelle); f. *Tarsius tarsier* (Ibnu Maryanto); g. Tarsius fuscus (Myron Shekelle);

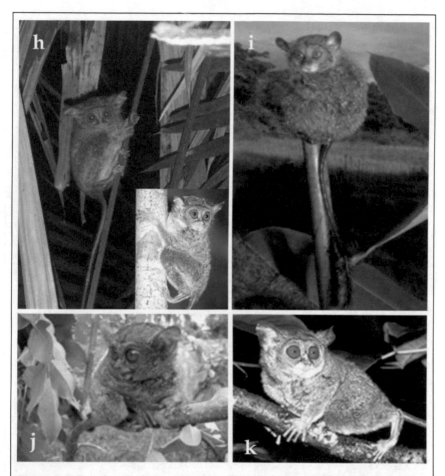

Figure 3. Indonesian Tarsier continued. h. *Tarsius supriatnai* (James Kumolontang & Lynn Clayton); i. *Tarsius dentatus* (Myron Shekelle); j. *Tarsius pelengensis* (M. Indrawan); k. *Tarsius sangirensis* (Myron Shekelle).

7. Selayar Tarsier

Tarsius tarsier

Other common names: Tarsius Selayar

Identification

This species is the smallest of all Tarsiers weighing approximately 0.120 kg. The head and body length is around 11 cm and the tail is naked and long (13.5 – 27.5 cm) with fine hairs on the tip. Possessing large eyes and ears, the head is round, the neck short, and the legs are long for leaping. The fur is dense and short and soft. The body color is mostly brownish-red with underlying mouse-grey tones. The eyelids, cheeks, submental region, rump, and lateral aspect of thighs in the adult, and more especially in the female, are rust-red. The flexor surface of the thighs and the ventral aspect of the base of the tail are whitish-yellow. The ventral surface of the thorax and abdomen is covered with grey hairs tipped with white. There is a black patch over the anterior part of the maxillae at the side of the nose. The throat is pale yellow. The ears are dark reddish-brown, thin, membranous and nearly naked. The forelimbs are short and the cheek teeth are adapted for a diet of insects.

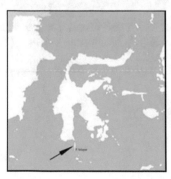

Geographic Range

This species is only found in Selayar island, south of Sulawesi.

Behavior and Ecology

Food mainly consists of insects, though small lizards, and crusta ceans are also reported to be taken. Their social organization is highly developed with pair bonds forming within larger social communities. Monogamous families, which consist of males, females and young, sleep in groups of up to 5 animals. They usually move by clinging vertically, then leaping quadrupedally, and climbing. This animal can leap an average of 1.4 m and a maximum of 5 - 6 m (Shekelle, 1997) .

Tarsius tarsier is completely nocturnal. It lives in the canopy and on the ground (0 - 9 m). They are notable acrobats in trees and shrubs. They are not believed to build nests. They usually spend the day sleeping on a hole in a big tree. Adults, juveniles and mature offspring regularly sleep together at the same site and sometimes can be found on the ground, where they are assumed to hide among the roots of trees. They mark their territories with urine left on bark. Physical contact is common and includes allogrooming, playing, tail inter-twining, and huddling. They may forage close together and cooperate in the defense of their territory. Little detail is known of their vocalization behavior but tarsiers have elaborate systems of acoustic communication, with up to 14 different types of call. Members of a family can often be heard calling back and forth to each other as they search for food. When they come out of or enter their sleeping tree a loud high-pitched noise can be heard repeatedly. It is similar to the sound of a rat squealing, but harder and deeper. It can be heard as far as 100 - 200 meters away in the forest. Females and males produce back and forth squealing noises at dusk. These sounds are different between females and males. As well as communication tools these sounds help to mark their territory (Gursky, 1997).

Conservation Status

The Indonesian government has protected this animal by law since 1931. Today the species is managed under the Indonesian Ministry of Environment and Forestry. Habitat destruction is currently the main threat to their survival. They are considered Vulnerable by the IUCN

Red list based on habitat loss alone. About 15-26% of their forest habitat was converted to agriculture between 1990 and 2000 and since that time at least an additional 10% has been lost.

Where to See It

Lowland secondary forest, scrub jungle and coastal scrub or mangroves are major habitats of this animal. You can fly from Jakarta to Makassar then go by car to Bira at Bontobahari, a small port in the eastern part of South Sulawesi and then to Parmatata on Selayar Island. The ferry will take 2 hours and the price is around USD5. You can also fly to Selayar from Makassar airport, which takes about 50 minutes. Many tourists come to Selayar for snorkeling or diving at Takabonarate Marine National Park.

8. Makasar Tarsier

Tarsius fuscus

Other common names: Tarsius makassar (Indonesian), Balao cengke (Patunuang-Makasar), Congkali (Malenreng), Pa'cui (Balang lohe).

Identification

The Makasar Tarsier has a notable tail which gradually widens toward the tip. Pelage color generally is rufous brown above and creamy below. They have black spots on either side of the snout and a white patch behind their ears. Head and body length is around 12.4 – 12.8 cm with a long tail relative to their body size of 24 – 26 cm. The weight of males is around 0.126 – 0. 133 kg while of females is around 0.113 – 0.124 kg.

Geographic Range

This species is found in the Southwest peninsular of Sulawesi. It is found in many different habitats including primary forest, secondary growth, community gardens, riparian forest from lowland to highland (> 600 m), for example, in Bantimurung Bulusaraung National Park in South Sulawesi (Shagir et al., 2011). While in Pangkejene Kepulauan, it was found up to 1200 m asl (Chaeril et al., 2011).

Behavior and Ecology

They are known to be arboreal and active during the night. Individuals are also assumed to return to one or more regular sleeping sites. There is currently no adequate information on the behavior and ecology of the Makassar Tarsier but presumably, like other tarsiers, their diet consist of insects and small vertebrates.

Sleeping trees are varied but mostly bamboo trees, fig trees or in palm trees (*Arenga pinnata*) or occasionally in rubber trees (*Hevea tiliaceaus)* with nests height up to 20 m above the ground. In the primary forest, however, they selected fig trees (*Ficus* spp) or bamboos (Sinaga, et al., 2009). It was also found in the tree of the karst mountain at Tompobulu village, Pangkep district at the fig trees with many liana around them and also found at the sugar palm (*Arenga Pinnata*) (Mansyur, 2012) .

Population densities of Makasar tarsiers in the resort Baloci of the Bulu Saraung National Park varied with 151 individuals/km^2 in secondary forest, 36 individuals/km^2 in plantations and 23 individuals/km^2 in vegetation near settlements (Mustari et al., 2013).

In other places, at the same park, its population density were between 13-66.8 animals/km2 with 2-6 individuals per family (Shagir, et al., 2011). While in Pangkep district, it was found 190 individuals of 67 families with an average of 2 - 4 individuals of each family (Chaeril, et al., 2011).

Conservation Status

The Makassar tarsier is listed under CITES Appendix II but its status on the IUCN Redlist has not yet been available as it is currently being assessed as a new species. However, as has been stated in the IUCN redlist in 2011 its previous taxon was categorized as Vulnerable.

Where to See It

They live in primary and secondary forest, thorn scrub, mangroves and montain forest. They are also found in proximity to urban areas, in crops and plantations. In Batimurung Bulusaraung National Park, this species has been observed to live in steep karst hills and to use small holes and interconnected tubes within the limestone. The Bantimurung Bulusaraung National Park is only 40-60 minutes drive from Makassar, South Sulawesi.

9. Sulawesi Mountain Tarsier

Tarsius pumilus

Other common names: Tangkasi gunung (Indonesian)

Identification

Tarsius pumilus is a small bodied tarsier with head and body length around 9.7 cm, a tail 20 -

21 cm long and weighing 0.481 – 0.501 kg for males and 0.52 – 0.575 kg for females. The tail is relatively short and naked except for the tuft which widens towards the tip. The dorsal pelage is reddish brown. The fur is silky and similar to that of *T. sangirensis*. The face is red with small ears and buff-colored spots behind them.

Geographic Range

Known from only three museum specimens collected over the course of the past ninety years, it is one of the most mysterious primate species. Inferences drawn from these specimens are that it is a small tarsier adapted for life in the mossy montane forests of Sulawesi at elevations of 1,800–2,200 m. Mountain tarsiers occur in South and Central Sulawesi. Their distribution is fragmented on isolated mountain tops of Lompobatang in South Sulawesi and Mt. Lorekatimbu and Rano Rano in Central Sulawesi.

Behavior and Ecology

Shekelle (2013) described the history of the discovery and description of this species. Musser and Dagosto (1987) were able to identify *T. pumilus* as a montane endemic tarsier adapted to the unique characteristics of the moss forest. They noted several morphological peculiarities shared by the specimens collected from Rano Rano and Rantemario. Most obviously, both specimens were quite small, although clearly adult. Linear measurements averaged about 75% of those seen in other Eastern tarsiers (those from the Sulawesi biogeographic region). Other similarities included the keeled, claw-like nails on the fingers and toes, which were very long, extending beyond the digital pad. The central lower incisors were relatively long, and scanning electron microscopy revealed fine striations, consistent with their having been used to comb their fur. The mountain tarsiers from Rano Rano and Rantemario were different from all others in having rather longer, silkier fur. Additionally, both exhibited enlarged auditory bullae.

After Musser and Dagosto (1987) described these specimens many primatologists searched unsuccessfully for live animals (Shekelle, 2008). It seems that this species did not duet, or produce other vocalizations that are common to other eastern tarsiers (Shekelle, 2008) making them more difficult to locate. In May 2000, a field assistant working for a small mammal survey of Lore Lindu National Park inadvertently trapped and killed the third known specimen at 2,200 m above sea level (Maryanto and Yani, 2004).

As with nearly every known tarsier acoustic form, vocalizations are absolutely diagnostic of Tarsius wallacei, even with relatively simple analyses. The human ear can easily diagnose the vocalizations of the new species from all other tarsier acoustic forms with minimal training. Similarly, visual inspection of spectrograms is sufficient to differentiate the duet call of *Tarsius wallacei* (Figs. 6, 7, and 8) from all its congeners, particularly as regards female phrases. Shekelle (2008a) described the duet call of Tarsius wallacei.

The first pygmy tarsiers seen alive since the 1920s were found by a research team led by Dr. Sharon Gursky and doctoral student, Nanda Grow from Texas A&M University, on Mount Rore Katimbo in Lore Lindu National Park in August 2008. The two males and a single female (a fourth escaped) were captured using nets, and were radio collared to track their movements. As the first live pygmy tarsiers seen in 80-plus years, these captures dispelled the belief among some primatologists that the species was extinct.

This species is assumed to eat animal prey such as insects and lizards, but further information on their ecology and behavior is unavailable.

Conservation Status

They are listed in CITES Appendix II and classified as endangered based on IUCN criteria B1 ab, which means that their distribution is highly restricted, in this case to the high tops, of 3 mountains Lompobatang, Lorekatimbu and Rano Rano in Sulawesi. As for other tarsiers, this species is protected under law No. 5, 1990. Habitat loss due to illegal logging and encroachment threatens this species in the wild.

Where to See It

They can be seen in mountainous regions of South and Central Sulawesi at elevations of 1800 - 2200 m. The Sulawesi mountain tarsier can be found on Rore Katimbo in Lore Lindu National Park in Central Sulawesi and Mt. Latimojong in South Sulawesi. To reach Rore Katimbu, you need to go to Palu (the capital city of Central Sulawesi) and rent a car to drive to the Lore Lindu National Park, which will take 3 - 4 hours.

10. Lariang Tarsier

Tarsius lariang

Other common names: Tarsius lariang (Indonesian), Ngasi (Central Sulawesi)

Identification

The Lariang Tarsier is different to all other Sulawesi maindland tarsiers. It is the largest, with an average head and body length of 12.1 cm and an average tail length of around 21 cm in males and 11.8 cm in females. The pelage has a very dark gray-buff coloration with a very dark thick terminal tail tuft and the thighs lack brown tones. There are distinct thick black paranasal stripes and well-marked black eye rims. The hair near the mouth is whitish with a bare spot at the base of each ear. Another distinctive feature is a long third digit (Merker et al., 2006).

Geographic Range

Tarsius lariang occupy western central Sulawesi, in the Lariang catchment near the confluence of its tributary, the Meweh River, and extends as far as north as Gimpu. The precise distribution has not been determined yet but the species is known to be parapatric with *T. Dentatus* (Merker et al., 2006).

Behavior and Ecology

This species mainly consumes insects, along with some small vertebrates such as frogs and lizards. Further information on its ecology and behavior is not available yet.

Conservation Status

Habitat loss and illegal logging is the major threat to this species. It also suffers from predation by dogs and cats, and the illegal pet trade. Currently, it is listed under CITES Appendix II and as Data Deficient in the IUCN Redlist. Further surveys are needed to determine its distribution and status.

Where to See It

The Lariang can be seen in secondary and primary forest, mangrove forest and gardens within forest landscapes. It can be seen in the forest near Gimpu, Central Sulawesi. You can fly from Jakarta to Palu city, the capital of Central Sulawesi. From there you can go by bus to Gimpu, which will take 4-5 hours depending on the condition of the road. You can also rent a car from Palu, which will likely take less time.

11. Wallace's Tarsier

Tarsius wallacei

Other common names: Tangkasi Wallace (Indonesian), Ngasi (Central Sulawesi)

Identification

The males of Wallace's Tarsier are slightly larger than the females. Males weigh around 0.1 – 0.124 kg while females weigh around 0.084 – 0.116 kg. The average head and body length is around 11.3 – 12.4 cm and tail length from 23.6 – 26.6 cm, which is dark with a long tail tuft. The pelage color is mottled yellowish grey with an off-white ventrum. Mottling is due to gray undercoat and scattered patches of light-gray black-tipped hairs. Wallace's Tarsier has a copper-colored throat that is clearly visible. There are distinct yellow to copper-colored eye patches above and below the eye. This species can be distinguished from its congeners via a characteristic duet song and its yellow-brown pelage coloration and a copper-colored throat. Genetic analyses prove Y-chromosomal and mitochondrial DNA sequences and also microsatellite allele frequencies to be absolutely diagnostic (Merker et al., 2010).

Geographic Range

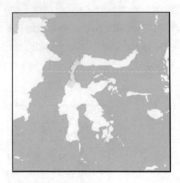

This species occupies a discontinuous range of forest in the province of Central Sulawesi. The northern and southern populations are isloated from each other. The northern population occurs just west of Tomini village in the northeastern part of its range and between Ampibabo and Marantale villages in the southern part of its range (Merker et al., 2010). The northern population borders the as yet undescribed form known as "Sejoli form" (Shekelle et al., 1997). The southern population is adjacent to *T. dentatus.*

Behavior and Ecology

Wallace's Tarsier is known to be a nocturnal, arboreal species. Presumably, they consume mainly animal prey, especially insects. Little is known of its behavior and ecology but it is assumed to be similar to that of other Sulawesi tarsiers.

Conservation Status

Within their estimated distribution range (3,150 km^2) the forest is mostly degraded and mountainous. The range of the southern population is tiny (estimated around 50 km^2). They are listed in CITES Appendix II and as Data Deficient in the IUCN Redlist. Clearly, further research is needed to confirm their distribution and status.

Where to See It

Wallace's Tarsier can be found near the city of Tinombo, north of the Isthmus Palu. They occur in the forest but can also be found in highly degraded forests and areas that have been cleared for agricultural purposes with mixed agroforestry and secondary habitat. You can fly from Jakarta to Palu city, which takes approximately 3 hrs. From Palu, you can take a bus to Tinombo or rent a car, which takes approximately 1-2 hrs. From Tinombo, you can go to see it in Gunung Sojol Nature Reserve.

12. Dian's Tarsier

Tarsius dentatus

Other common names: Ngasi (Central Sulawesi)

Identification

This species closely resembles the spectral, lesser, and western tarsiers, anatomically, morphologically and in coloration. Generally, head and body length is around 8.5 -16 cm, tail length is 13.5 - 27.5 cm and adult weight is 0.080 - 0.165 kg. The tail is naked, long and has a few short hairs on the tip. The eyes and ears are remarkable large, the head is round and the neck short. The head can be rotated nearly 360 degrees. The body coloration ranges from greyish brown to dark brown and the under parts are buff or slate.

Geographic Range

This species occurs in the eastern portion of the central core of Sulawesi, Indonesia.

Behavior and Ecology

Dian's tarsiers are commonly known to be insectivorous but they also prey on other small animals, including frogs, lizards, and small mammals. They move by leaping, walking, hopping and climbing. They can make quick leaps of several meters. Merker & Muehlenberg (2000) found that the decrease of tress for locomotor support and potential sleeping tree abundance as well as the noise resulting from logging, was the cause of the low population densities they encountered in logging areas. They also found that the density of small trees and branches was significantly less in primary forests than in three other habitat types, which suggests that substrate abundance is not a major factor influencing distribution patterns. The highest population density of *T. dentatus* was found in the habitat with the smallest abundance of vertical and horizontal locomotor supports.

Knowledge of their behavior and ecology remains poor. They are nocturnal and generally arboreal but sometimes they can be seen on the ground searching for food. On any flat surface they also can leap frog-like and they can walk on all fours with the tail hanging down. They are monogamous, with a small territory that is actively defended. The territory is marked with urine and the scent of various glands.

Sleeping sites vary with habitat type. In primary forest, they often sleep in tree cavities, mainly of strangler figs (*Ficus* spp.). In secondary forest, they usually sleep in tree cavities, bamboo stands, or shrubs (MacKinnon & MacKinnon 1980, Tremble et al., 1993). Groups of *T. dentatus* have been observed usually returning to their initial sleeping sites each morning, although there is a tendency to change to alternate sites when disturbed.

The population density of this species has been found to vary from 22-250 individuals/km2 (Gursky 1998) and 45-268 individuals/km2 (Merker, 2003). The density is correlated with food and available trees for nesting. For example, Fig trees, which are commonly used for nests can also supply food as their insect prey are attracted to figs (Merker et al., 2010). Merker et al., (2005) reported that the population density of this tarsier in the disturbed habitat in Lore Lindu National Park was 45 individuals/km², while in the undisturbed forest the density can reach 268 individuals/km².This species is Vulnerable in the IUCN Red List due to habitat loss. At least 30% of the habitat of this species has been converted to non-forest in the past 20 years.

The voice of males and females of *T. dentatus* is still relatively abundant in the observed area. This species can be found not only in primary forest, but also in secondary habitats and in forest with logging and plantations. The habitat types we investigated, with varying human activities, support different population densities of *T. dentatus*. "Primary forest", as the least disturbed habitat, supports a significantly higher relative population density than the three more disturbed habitat types in focus. Merker & Muehlenberg (2000) found the highest density of *T. dentatus* in forest with interspersed small plantations, which contained even slightly more tarsier groups than the same area of primary habitat

Conservation Status

From 1990 to 2000, from 15 to 26% of the forest habitat on Sulawesi was converted to agriculture, and since that time at least an additional 10% has been lost. Conservation strategies for this unique and endemic species should consider the different effects of varying human activities. Tarsiers avoided fallows, maize fields and rice paddies but strongly selected dense shrubbery for shelter and for travelling through (Merker and Yustian, 2008). Although some land-uses seem to be less detrimental to tarsiers than others, conserving intact forest is still the best bet. More detailed research is needed to assess the impacts of human activities on tarsiers.

Where to See It

This species can be seen in Lore Lindu National Park, which is 70 km from Palu, Central Sulawesi. In Kamarora, Lore Lindu NP, this tarsier is being studied by both foreign and local researchers. Lore Lindu NP has two entry points, through Palu and Tentena (Danau Poso).

- Palu ⟶ Sidoanto (65 km) you can get to Sidoanto in around 2 hours by car and continue by walking, horse riding, or rental motor bike to Danau Lindu (20 km).
- Palu ⟶ Kamarora (50 km) can use public transportation, which will take about 2.5 hours. Kamarora is a great place to observe tangkasi. Accommodation is available and there is a hot spring.

13. Peleng Tarsier

Tarsius pelengensis

Other common names: Tangkasi peleng (Indonesian), Lakasinding (West Peleng), Siling (East Peleng)

Identification

This species is only known from museum specimens. It is relatively large with a head and body length of 12 – 14 cm and a tail length of 25–27 cm. It has a dark brown coat with contrasting creamy-tipped thighs and a vague black spot on its nose.

Geographic Range

Presumably, *T. pelengensis* can still be found on Peleng Island, where the museum specimens were obtained, off the coast of east peninsula Sulawesi.

Behavior and Ecology

No comprehensive data is available on the Peleng Tarsier's ecology and behavior, but they are assumed to consume mainly invertebrates such as moths and crickets, and small vertebrates such as frogs and lizards.

Conservation Status

The species is listed in CITES Appendix II and classified as Endangered in the IUCN Redlist. It is mainly threatened by habitat loss. It is endemic to a small island of approximately 2,406 km2 without any protected areas. Less than 10% of the island is considered to be suitable habitat for this species. Several researchers used a mathematical model to calculate the population and came up with 156 individuals/km2 (Gursky, 1997).

Where to See It

It can be expected to be seen in the primary and secondary lowland forest of Peleng Island. To reach Peleng Island you can go through Palu, Central Sulawesi. You can then go by bus to Luwuk, which will take between 12 and 20 hours, or by small plane, which will take about one hour. From Luwuk, there is a ferry or you can rent a motor boat to Peleng Island if the weather is good. It is probably better to explore the forest away from local settlements when looking for this species.

14. Great Sangihe Tarsier

Tarsius sangirensis

Other common names: Tangkasi sangir (Indonesian), Senggasi, Higo, Tenggahe (Sangir)

Identification

Tarsius sangirensis is similar in color to *T. tarsier*, which is yellowish grey. Its Body length is between 11.5 and 12.5 cm, with a tail almost twice the length of its body, between 22.5 and 24 cm. There is almost no hair at the tip of the tail. Body weight is around 0.11 to 0.12 kg.

Geographic Range

 This species of can only be found on Sangihe and Talaud Islands, North Sulawesi.

Behavior and Ecology

The Great Sangihe tarsier eats mostly large arthropods and some small vertebrates. They are nocturnal and arboreal, moving from one tree to another by jumping. When hunting for food, they stay still in branches and jump to catch prey that wanders by.

Similar to other tangkasi, Tangkasi Sangir form a monogamous family system. They live in small groupings of 2–6 individuals. In particularly disturbed habitat, they might sleep in dispersed social groups, which may be a response to the risk of predation. There has been no systematic study of the wild population of this species, but they were surveyed by Shekelle and associates (Shekelle et al., 1997, Shekelle, 2003) and Riley (2002). Riley found that the habitat preference of *T. sangirensis* is primary forest, but that it may occur in secondary habitats such as sago swamps, scrub, nutmeg plantations, coconut plantations, and secondary forest regrowth. These results are similar to those of Shekelle and colleagues, except that Shekelle did not encounter any in primary forest, and Shekelle and Salim (2009) indicate that it is very unlikely that any primary forest now remains on the islands.

Conservation Status

According to the IUCN red list, the conservation status of *Tarsius sangirensis* is Endangered as they only found two islands that have severely fragmented habitat with ongoing forest clearing, making them seriously threatened indeed. There is no single park that can protect this tarsier on either island.

Where to See It

They are commonly seen in bamboo and sago forest on Sangihe and Talaud Islands. There is a flight to these islands from Manado, but you can also take the fast ferry, which is 4 to 5 hours or the ordinary ferry, which takes 10 hours. Besides looking for this species, these islands offer dive resorts, endemic birds, and more than 70 smaller islands surrounding Sangir Talaud Islands.

15. Siau Island Tarsier

Tarsius tumpara

Other common names: Tangkasi Siau (Indonesian), Tumpara (Siau)

Identification

This species of tarsier is differentiated from other species by unique features of the tail tuft, pelage, skull, and by its vocalizations. *Tarsius tumpara* possesses a tail tuft that is characterized by short sparse fur that is light in coloration, and in this way resembles *T. sangirensis*. *T. tumpara* also resembles *T. sangirensis*, in that it exhibits reduced furriness of the tarsal and paralabial hair that is white and pronounced. The dorsal fur of *T. tumpara* lacks the golden brown coloration that distinguishes *T. sangirensis*. The mottled brown ventral surface and dark gray undercoat of *T. tumpara* contrasts with the golden brown ventral surface of *T. sangirensis* and its nearly white undercoat. The skull of *T. tumpara* is larger than three of the four available skulls of *T. sangirensis*, but relatively narrow across the eyes.

Geographic Range

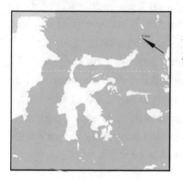

It is found only on Siau island, approximately 200 km north of North Sulawesi (Shekelle et al., 2008).

Behavior and Ecology

Knowledge of this species' behavior and ecology remains limited. It is known to be an insectivore. It is arboreal and nocturnal. Similar to other tarsiers, it is likely to be monogamous. It occurs from lowland wet forest up to mountain forest.

Conservation Status

The Siau island tarsier is listed as Critically Endangered in the IUCN Redlist, which means there is a very high risk of extinction. There is little available habitat and the species is vulnerable to any changes in that habitat.

Where to See It

To see this tarsier you need to go to Siau Island. From Jakarta you can fly to Manado, North Sulawesi. You can then take the Bahari express ferry or the Oasis jet boat to Siau Island. This trip will take around 4 hours in a comfortable boat. There are three classes of ticket, which range from Rp. 160,000 to 290,000. Some ferries are also available during the evening.

Siau Island is an island in Kepulauan Siau Tagulandang Biaro (SITARO) Regency, North Sulawesi Province. Karangetan Mountain, also known as Api Siau Mountain is the prominent geographical feature. This volcano has erupted more than 40 times since 1675, most recently in 2015, although it posed no danger to people. Siau Island is also famous for its nutmeg, which is regarded as one of the best in Indonesia. There are several restaurants that stay open until late, especially near the market area. Siau also has the Lehi hot spring at Ondong, which is located in the west side of the island.

16. Supriatna's Tarsier

Tarsius supriatnai

Other common name: Bumbulan, Mimito (Gorontalo)

Identification

Supriatna's tarsier was discovered and described by Myron Shekelle and friends in 2017.

According to Shekelle et al., (2017) this tarsier has a larger bare spot at the base of the ear, a less shortened hind foot, a very long tail, and longer middle fingers than other species. From the small sample analyzed body weight and tail length are probably within the range of several other tarsiers: body weight (female = 0.104 to0. 114 kg, and males = 0.135 kg, unfortunately the sample size is very small); tail length (female = 2.32 to 2.43 cm, and male = 2.46 cm (Shekelle, 2003). Museum specimens indicated that a species with skull and teeth very similar to *T. spectrumgurskyae n. sp.*, but the two specimens that were measured, compared to nine of the latter, have a lower anterior central incisor, and larger first and second molars. Genetic analysis identified a *T. spectrumgurskyae - T. supriatnai* clade, which is different from all others and a difference between the two species. Driller et al., (2015) in Shekelle et al., (2017) estimated a divergence date of 0.3 MYA for the separation of the two.

Geographic range

This tarsier distributed in Northern Sulawesi from the Isthmus of Gorontalo westward at least as far as Sejoli, and probably reaching up to Ogatemuku (see Driller et al., 2015), but not as far as Tinombo. In the west, this species borders with *Tarsius wallacei* and in the east with *Tarsius spectrumgurskyae*.

Behavior and Ecology

MacKinnon and MacKinnon (1980) originally described a Gorontalo form of a tarsier population that was referred to as the Libuo form in various papers by Shekelle (Shekelle et al., 1997; Shekelle 2003, 2008). Myron Shekelle and his colleagues (2017) studied this tarsier carefully in Nantu forest in Gorontalo and found the acoustic duet of this form was characterized by a ~2 to 5-note female phrase accompanied by male calls, which was very different from other species

Conservation Status

Since this tarsier was only described in 2017, it should be categorized as Data Deficiency in the IUCN red list. However, it is highly likely to be vulnerable, along with most tarsiers in Indonesia.

Where to see it

Nantu Protected area is the best place to see this tarsier. There is a research station within Nantu, close to Paguyaman district. Dr Lynn Clayton has been studying mammals at this station and you can also see other animals, such as the unique Babyrousa (*Babyrousa babyrousa*), the endemic animal that looks like both a pig and a deer. The males have long tusks growing upwards and outwards. There are also several endemic Gorontalo black Macaques (*Macaca hecki*) and the dwarf buffalo or lowland Anoa (*Anoa depresicornis*).

In order to go there, you need to fly from Jakarta to Gorontalo, which takes approximately 3 hours. From Gorontalo city, you may take a bus to Paguyaman and then from there you may charter a car to the research station that may take 3-4 hours.

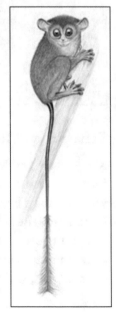

17. Gursky's Spectral Tarsier

Tarsius spectrumgurskyae

Other common names: Tangkasi, Wusing

Identification

The original name of this species is *Tarsius spectrum*. Because of the major revision of tarsiers in the northern Sulawesi area, then Shekelle et al., (2017) named this species in honor of Dr. Sharon Gursky, who has studied the behavior and ecology of this species for many years in Tangkoko Batuangus Protected Area, North Sulawesi. Shekelle et al., (2008, 2010) found their small genetic data set to be broadly consistent with the hypothesis that acoustic forms are distinct species, but did not find the Manado form to constitute a single monophyletic clade. More recently, Driller et al., (2015), using more genetic evidence, found support for the separation of *T. spectrumgurskyae* from the rest of tarsiers and dated its divergence at 0.3 MYA.

Surveys of wild populations indicate that body weight and tail length are within the range of several other species of tarsier: body weight (female 0.95 to 0.119 kg, n = 24; male 0.104 to 0.126 kg, n = 11), tail length (female = 21. 3 to 26.8 cm, n = 22; male = 22. 0 to 25.8 cm, n = 9) (data from Shekelle, 2003).

Geographic Range

It can be found in the Northern Sulawesi, bordering in the west with *T. supriatnai.*

Behavior and Ecology

The Manado form was originally described by MacKinnon and MacKinnon (1980), and further examined by Niemitz et al., (1991), Shekelle et al., (1997), Nietsch and Kopp (1998), Nietsch (1999), and Shekelle (2003, 2008). More recently, Yi et al., (2014) found that this species and *T. supriatnai* are easily separable with quantitative analysis.

There has been a great deal of information on this species. Long term studies have been conducted in Tangkoko Batuangus Protected Area by John and Kathy Mackinnon, scientists from UK, who began studying this species in the Tangkoko Batuangus Protected Area in the early 1980s, followed by Tim O'Brien and Margareth Kinnard and Sharon Gursky respectively. This tarsier is the best-known tarsier species in the world because it has been studied for the last three decades. The behavior and ecology of the species have been filmed and shown on the National Geographic Television Channel.

Conservation Status

This tarsier has been categorized as vulnerable on the IUCN Red list.

Where to see it

They can be seen easily in Tangkoko Batuangus Protected Area, approximately 60 km to the east of Manado, North Sumatra. Besides the research station, there are also other facilities such as local guides who can speak English and have a great deal of knowledge on their behavior and ecology. Local home stays can be found at Batu Putih village, near the protected area. Good and luxury hotels can be found in Manado. Rental cars are available in the airport at Manado, North Sulawesi.

18. Western Tarsier

Cephalopachus bancanus

Other common names: Sumatra: Singapuar (Bengkulu); Krabuku (Lampung), Palele (Belitung), Mentilin Ingkir, ingkit, Beruk-puar (Bangka), Kalimantan: Ingkir, Linseng (Ngaju), Ingkat (Iban); Page (Tidung), Makikebuku (Karimata), Singaholeh (Kutai), Tempiling (Kalbar), Binatang hantu, Simpalili (Melayu).

Identification

In general, head and body length is around 11.4 - 13.2 cm and weight around 0.11 - 0.138 kg for males and 0.1 - 0.19 kg for females. Compared to all other tarsiers, the Western Tarsiers have relatively larger eyes, shorter ears and longer hindlimbs and hands. The skull is broader as it has heavily flared eye sockets.

- *C. b. borneanus* has a dark gray and rufous brown dorsal pelage, which is typical of many tarsiers.
- *C. b. bancanus* has notably yellow ochre tones in the pelage that are unseen in other tarsiers. There is a dark spot on each knee and the facial mask is less vivid than in the eastern tarsiers.
- There are few comparative descriptions of the *C. b. saltator* and *C. b. natunensis* subspecies. It is possible that the first is a synonym of *C. b. bancanus* and the second a synonym of *C. b. borneanus*.

Geographic Range

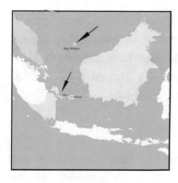

This species has 4 subspecies:
- *C. b. bancanus*

This subspecies is found in South Sumatera. Precise distribution on Sumatera Island is unknown, but appears to be restricted in the southern Sumatra, bordered by the Musi River.
- *C. b. borneanus*

This subspecies occurs on Borneo and Karimata Islands

- *C. b. natunensis*

This subspecies occurs on Serasa Island and possibly nearby Subi Island in the South Natuna islands, but not in the North Natuna Islands.

- *C. b. saltator*
  This subspecies is limited to Bangka Belitung Islands.

Behavior and Ecology

The Western tarsier is known as a carnivorous species, which consumes only live animal prey such as insects, freshwater crabs, frogs, lizards, birds, bats, and even venomous snakes such as the banded Malaysian coral snake (*Calliophis intestinalis*). Similar to other tarsier species, they are nocturnal and arboreal. They become active right before sunset and cease around sunrise. Activities peak in the early evening and just before sunrise, which is predicted to be associated with feeding. Females have exclusive home ranges that overlap with one or more other individuals. Apart from courtship and mating, direct contact is rare.

Conservation Status

Western Tarsiers are listed under CITES Appendix II and as vulnerable in the IUCN Redlist. This species is threatened by the loss of habitat due to forest conversion to palm oil plantation, fires, logging and also hunting for the illegal pet trade, particularly around Way Kambas National Park and in the entire Lampung Province.

Where to See It

They are found in almost all forest or partly forested habitats except for urban areas and intensive agricultural areas where there are no potential sleeping sites or where pesticides are commonly used. Nonetheless, they are also found on forest edges and in plantations. They usually occupy elevations below 100 m. Recently, a tarsier was caught at 1200 m asl which suggests that it also occurs in some highland areas.

C. FAMILY CERCOPITHECIDAE (Gray, 1821)

1. Subfamily: Cercopithecinae

There are twelve primate genera in the subfamily Cercopithecinae. All of this subfamily lives in Asia and Africa. One of the genera *Macaca* is distributed widely in Asia with only 1 species found outside Asia, the Barbary macaque in Africa. The Barbary macaque is mainly found in the atlas and rift mountain ranges of Morocco and Algeria.

Macaca: Macaques

There are 22 species spread throughout Asia from the savannah in India to the tropical rain forests of South, Southeast Asia and East Asia plus the snow-covered mountains of Japan. In Indonesia there are 10 species, also widely distributed, from Sumatra, Kalimantan, Java, and Sulawesi to the islands of Nusa Tenggara. The long-tail macaque (*Macaca fascicularis*), has the widest distribution of all Indonesian species. Besides Indonesia, long-tail macaques can be found in Laos, Vietnam, Cambodia, Thailand, Malaysia, and the Philippines. In contrast, the Beruk (*Macaca nemestrina*) can only be found on the Malaysian Peninsula and in Sumatra and Kalimantan. Some species are endemic to Sulawesi and the Mentawai islands.

Body measurements vary greatly between species. The smallest is the long-tail macaque, whose mature body weight is approximately 3 kg. The largest is the Sulawesi black macaque (*Macaca tonkeana*) that attains 15 kg. The gestation period is between 153 and 179 days, and most give birth to one offspring.

© The Author(s), under exclusive license to Springer Nature Switzerland AG 2022
J. Supriatna, *Field Guide to the Primates of Indonesia*,
https://doi.org/10.1007/978-3-030-83206-3_6

Figure 5. Indonesian macaques: a. *Macaca fascicularis* (Anton Ario & Jatna Supriatna); b. *Macaca nigrescens* (Noel Rowe & WCS-IP); c. *Macaca nigra* (Jatna Supriatna); d. *Macaca nemestrina* (Jatna Supriatna); e. *Macaca siberu* (Noel Rowe); f. *Macaca pagensis* (Noel Rowe); g. *Macaca maura* (Jatna Supriatna); h. *Macaca tonkeana* (Noel Rowe); i. *Macaca ochreata* (Putu Sutarya); j. *Macaca hecki* (Michel Gunther).

19. Long-tail Macaque

Macaca fascicularis

Other common names: Sumatra: Cigaq (Minangkabau), Karau, Cigah, Warik; Kalimantan: Ambuk (Dayak), Jibalau (Murut), Bakey (Beaju Dayak), Kode (Kutai), Bakei, Bangkoi (Ngaju), Warek (Dusun); Java: Ketek, Bedes (Tengger), Bojak, Mondang, Munyuk, Kenyung; Kunyuk (Sunda); Madura: Motak, Ketang; Timor: Belo, Slai.

Identification

The head and body length is from 37 to 63 cm in males and from 31.5 to 54.5 cm in females. They have noticeable long tails (male 36 to 71.5 cm, female 31.5 to 63.8 cm), which is presumably an adaptation to their arboreal forest habitat. *M. fascicularis* is the smallest Indonesian macaque, with average adult male weight between 3.4 and 12 kg and female weight between 2.4 and 5.4 kg.

The dorsal pelage coloration varies geographically, from buff to yellowish-gray; golden brown to reddish brown; dark brown to blackish. Crown hair is directed backward and outward and is more brightly colored than the hair on the back of the head. Infants are born black with bare pinkish facial skin. Infant body color lightens within a few months to gray or gray-brown, except for the pate, which may remain dark for many months. The face of adults is gray or tawny, in contrast to the eyelids and the skin above them, which are lighter in color. Individuals also vary in facial ornaments such as mustaches and cheek tufts. Adult males often have cheek whiskers and a mustache. Adult females also display a large facial beard that extends around the chin. Some individuals have a dark cheek stripe below each eye. In general, individuals are larger in the coastal or lowland regions and smaller at higher elevations. Estrus, occurring at puberty, is characterized by swelling on the bare skin between the root of the tail and the anus. The skin around the vulva frequently reddens. This swelling becomes less conspicuous and varies in size and duration as the individual ages (Mittermeier et al., 2013, Engelhardt et al., 2005)

Geographic Range

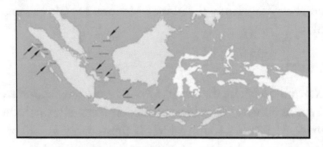

Indonesia has 5 subspecies of long tail macaque. They are as follows:
- *M. f. fascicularis*, occurring in Greater Sunda Islands (Sumatera, Borneo, Java, Bali, Sumba, and Timor)
- *M. f. fuscus*, found only on Simaleu Island off the northwest coast of Sumatera
- *M. f. karimundjawae*, found on the Karimunjawa Islands and presumably nearby Kemujan, Central Java
- *M. f. lasiae*, found only on Lasia Island, Aceh Province.
- *M. f. tua*,which is found only on Maratua Island, West Kalimantan.

Behavior and Ecology

This species lives in primary and secondary forest, from the lowlands to highlands more than 1000 m above sea level. In the highlands, they are usually found in secondary forest or agricultural areas. They usually choose trees along rivers to rest in. *M. fasicularis* is a frugivore - omnivore and is known to be an opportunistic feeder.

In the mangrove swamps they have been observed eating the fruit of *Sonneratia spp., Avicennia spp.,* and the nipa palm (*Nypa fruticans*). In the agricultural areas, they eat rubber fruit, rice shoots, and young corn plants. In secondary and riverine forests, however, they consume mostly fruit, leaves, flowers, some large insect larvae, fungi, vines, and and sometimes they even eat clay in order to reduce toxicity from eating bark and sap. They are also known to feed on crabs and other crustaceans, shellfish, prawns and other littoral animals exposed by the tide.

Long tail macaques live in groups usually consisting of many males and females. The number of individuals varies. In mangrove forest the number is between 10 and 20 individuals and in primary forest, 20 to 30 individuals. The size of a group is determined by whether there are predators and/or food abundance in the area. Predation detection also may be a major determinant of the social behavior of this species. In other groups, the number of males and females is usually the same. Inter-male competition often occurs within groups. Group disputes often occur but when there is plenty of food, disputes are less likely.

M. fascicularis never moves bipedally on the ground. It is known as a swimmer when need be and usually swims doggy style across rivers. The home-range size is usually between 50 and 100 ha. It is speculated that they feed in the same areas throughout the year. The predators of this macaque are felids (tigers, leopards), raptors (hawks, eagles), and snakes (pythons). When encountering danger, they usually make a loud and shrieking sound. The presence of a group can be detected by the sound "krra!"

Conservation Status

This species is listed as Least Concern in IUCN Redlist. While some subspecies are endemic to small islands such as Karimunjawa, Lasia,

Simeuleu and Maratua, research on this species's status are still limited. The species is threatened by habitat degradation, especially in Java, and captured for experimentation and as pets. Since the 1770's, long tail macaques have been exported for biomedical purposes and for psychological research. The species is not yet seriously threatened, but it has lost more than 70% of its natural habitat, down from almost 218,000 km^2 to 73,400 km^2, so there is no room for complacency. The macaque is a pest in agricultural areas. It can destroy rice paddies, rubber seedlings, and fruit trees.

Where to See It

This species has the ability to adapt to many different habitat types. It can be found from mangrove forest to primary and secondary hill and mountain forest. It is also often found near human habitation in many places in Sumatra, Kalimantan, Java, Bali and even to Timor Leste. In Bali, it is commonly found near temples such as those in Sangeh and Ubud. A population of long-tailed macaque can even be found in Muara Angke mangrove forest in North Jakarta Bay.

20. Sunda Pig-tailed macaque

Macaca nemestrina

Other common names: Sumatra: Baruk, Berok, Baruak (Minangkabau), Bui (Aceh), Bodat (Tapanuli), Kalimantan: Empau (sea dayak), Basuk (Tagal), Gobuk (Nurut), Cerok (Kapuas), Bankui (Pleihari), Malaysia: Brok

Identification

Macaca nemestrina is a thick-set macaque with a short pig-like tail, hence the common English name of pig-tailed macaque. The body color is olive-brown with whitish under parts. The top of the head and neck is dark brown. Adult male pig-tails have elongated and often erect hairs on the upper back and shoulder. During the estrus period, females have obviously swollen, pink buttocks (Rowe, 1996).

Sexual dimorphism is seen with males being larger, 53.2-73.8 cm, than females, 43.4-57.6 cm, heavier, 6.2 to 14.5 kg, than females, 4.7 to 10.9 kg, and broader in the shoulder and chest. Canines in the males also much larger than in the females, 12 mm compared to 7.3 mm, and are used in aggressive interactions (Mittermeier et al., 2013).

Geographic Range

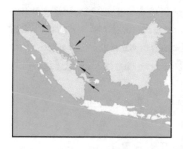

Sunda Pig-tailed macaques are widely distributed throughout Peninsular Malaysia, Thailand, the Malaysian states of Sarawak and Sabah on Borneo and Brunei Darussalam, also on Borneo. In Indonesia they occur throughout West, Central, South and East Kalimantan, Lampung, Bengkulu, South Sumatra, Jambi, West and North Sumatra, Aceh and Bangka Island. They are known to occupy riverine and swamp forests, farm lands and plantations, and secondary forest. However, they occur in the greatest density in upland primary rainforest.

Behavior and Ecology

Diet consists mainly of fruit, seeds, and arthropods. The majority is fruit with only 0.4% appearing to be ripe fruit. Other foods include nestling birds, termite eggs and larvae, and river crabs. Sometimes, they also consume the crops of field gardens and are therefore considered to be pests. The diet of this species in primary rainforest is qualitatively similar to that of *M. fascicularis* in that same habitat type. They are frequently found feeding on the ground (Rowe, 1996).

Pig-tailed macaques are diurnal and able to travel freely to forage in all levels of the rainforest. They are also well known to be semi-terrestrial. They move by quadrupedal climbing, terrestrial quadrupedal walking, and bipedal leaping and they can stand on two legs. Groups of *M. nemestrina* are thought to be far more wide ranging than groups of *M. fascicularis*, although diets appeared to be very similar. Home range size is estimated to be about 1 - 3 km². Juveniles and adult females are more often seen on branches, while adult males less so (Oi, 1990).

The social structure is multi-male/multi-female with distinct dominance hierarchies within each sex. Adult males are dominant over adult females and sometimes show aggression towards females at feeding sites. Females are capable of attacking low-ranking adult males with the help of female relatives, to defend food sources. They see low-ranking males as competitors. Females tend to stay with their natal group while males will emigrate after five or six years of age (Oi, 1996).

Conservation Status

This species is listed as vulnerable on the IUCN Red list. The primary threat to this species is habitat loss for oil palm plantations, which has long been an issue within their distribution range. Besides that, the species is considered to be a pest of crops so they are sometimes persecuted by farmers. In Peninsula Malaya farmers and plantation managers hunt them within plantations to reduce crop damage, pet trade, and for food (Nowak, 1999; Luskin et al., 2013). According to Richardson et al., (2008) the total population of *M. nemestrina* has declined by approximately thirty percent since the early 1970s.

Where to See It

In Sumatra, pig-tailed macaques can be found in forest areas, especially national parks such as Way Kambas, Bukit Barisan Selatan, Kerinci

Seblat, Bukit Tiga Puluh, Bukit Dua Belas, Tesso Nilo and Leuser and also nature reserves such as Seulawah and Janto. In Kalimantan they can be found in the national parks, Gunung Palung, Betung Kerimun, Kayan Mentarang, Kutai, Danau Sentarum and Sebangau as well as many other conservation areas.They are mostly found in primary and secondary forest in the lowlands and in hill forest up up to 1000 m a.s.l. There are no reports of them being seen in Mangroves.

21. Pagai Macaque

Macaca pagensis

Other common name: Siteut (Sipora, North and South Pagai)

Identification

Macaca pagensis similar features to *M. nemestrina*, except that the hairless tail and dorsal area are darker. Dorsal coloration is dark brown with contrasting pale ocher along the sides of the neck and the front of the shoulders and brown legs and reddish brown arms (Groves, 2001). In size and habits *M. pagensis* resembles *M. fascicularis* rather than *M. nemestrina*. Head and body length averages 53 cm with a 13 to 16 cm tail, while females are 43 to 46 cm in length with a 10 to 13 cm tail. The average body weight in males is between 6 and 9 kg and in females between 4.5 and 6 kg (Mittermeier et al., 2013).

Geographic Range

The Pagai macaque is restricted to the Mentawai Islands off western Sumatra (Siberut, Sipora, North and South Pagai). Normally this species may be found in all habitats except coastal mangroves. It is found in both primary and secondary lowland forests. Sometimes, it prefers local gardens and dense shrubby growth. It is more abundant in coastal swamp forest than inland forest where its prefered habitat is riverine forest.

Behavior and Ecology

The diet mainly consists of various types of fruit, seeds and young leaves. Besides plants, they also consume mollusc on the ground by rivers. This species is often seen in coconut plantations picking the fruit so they are considered to be agricultural pests by the locals. Only adult males are able to open the coconuts.

They spend most of their time in the lower levels of forests or on the ground. A group of *M. pagensis* can vary from 2 to 12 individuals in a polygamous arrangement consisting of one adult male and one or more adult females and their offspring. They are known to interact at the same food trees with two other colobines in the Mentawai Islands, the Mentawai Langur (*Presbytis potenziani*) and the Pig-tailed Langur (*Simias concolor*).

They seldom vocalize so little information is available on their vocalization behavior. Their repertoire is restricted to soft contact calls. Because there have been no long-term studies, little is known of the ecology or behavior of Pagai macaques, especially detailed distributional information and population size.

Conservation Status

The Pagai macaque is completely protected under Indonesian law, However, in the wild, it is threatened by hunting as local communities eat the meat and use parts of the animal in traditional ceremonies. Besides that, on Siberut, Sipora, North Pagai and South Pagai islands, they are also poisoned as they are considered to be agricultural pests. For example, 'Aldicarb' is a brand of poisonous liquid used by the Sikakap community to kill this macaque as well as monitor lizards (*Varanus* spp.). This has happened as forests have been converted to coconut plantations.

M. pagensis is listed as critically endangered on the IUCN Redlist. In 1986, it was listed as indeterminate. The species was determined to be endangered in 1988, and upgraded to critically Endangered in 1996 (IUCN, 2004). The most recent estimates suggest densities of 7-12 individuals/km in suitable habitat on the Pagai Islands (Paciulli, 2004). These ranges of variation in density estimates are related to habitat quality. Logging and hunting have decimated the population from approximately 15,000 in 1980 to only less than 3700 individuals (Mittermeier et al., 2013).

Where to See It

The best place to see this macaque is at the Betumonga Research Station, North Pagai. But you can also see this macaque in North Pagai, South Pagai and on Sipora islands. There is a ferry from Padang city of West Sumatra province to Sikakap, on North Pagai island. Recently the fast ferry, "Mentawai Fast", has begun running 3 days a week leaving Padang at 6 am and arriving at Sikakap at 10am. The regular ferry "Ambu Ambu" is an overnight journey, taking 13 hours, leaving Padang at 5 pm and arriving Sikakap at 6 am the following morning.

22. Siberut Macaque

Macaca siberu

Other common name: Bokoi (Siberut)

Identification

Macaca siberu was recognized as a distinct species by Kitchner and Groves (2002). Previously it was considered to be a subspecies of *Macaca pagensis*. This species is similar to *M. pagensis*, but has a broader face, a much darker black back and a white underside (Groves, 2001; Mittermeier et al., 2013). It also has white cheek patches while *M. pagensis* has black cheek patches. Neck patches are not found in *M. siberu*. Male body length is between 47 and 48 cm while female body length is between 40 and 45 cm (Mittermeier et al., 2013).

Geographic Range

This species is found only on Siberut Island, off the west coast of Sumatra.

Behavior and Ecology

Siberut macaques spend most of their time in dense continuous forest. Groups of 15 to less than 30 individuals have been observed (Abegg and Thierry, 2002). They utilize lower forest strata, from 0 to 10 m in height. Adult males spend more time in lower strata compared to adult females and juveniles.

The diet consists of fruit (75.7%), athropods (11.9%), mushrooms (4.5%), leaves (4.4%) and some others (pith, sap, shoots and flowers). This species has a high degree of frugivory compared to other macaques (Ritcher et al., 2013). Females have been observed to feed on pith from trunks that have already been opened. *M. siberu* have also been observed to occasionally catch and consume crabs and shrimp from rivers.

In a dense forest, female leaders make repeated sounds to keep in touch with the rest of her group. The call is a loud sound of 'Kof Kof Kof Kon Kon Kon Kon', which is returned by the other members of the group.

Conservation Status

This species is vulnerable on the IUCN Red list, and protected under CITES Appendix II. As *M. siberu* is a newly recognized species, conservation action in Indonesia has been slow in spite of the threats it faces. The Mentawai Islands suffer from habitat destruction due to logging and conversion to oil palm plantations which threatens the existence of this species. Around 60% of the Mentawai Island forest cover has been lost (Whittaker, 2006). With the decreasing availability of fruit that forest loss entails, this species is likely to raid crops, which brings it into conflict with local farmers. Local people usually hunt and trap them using ground traps baited with sago, where a whole group can be caught at once. Both habitat loss and hunting have resulted in a decline in population size so that now the number is predicted to be between 17,000 and 30,000, down from 39,000 in 1980 (Whittaker, 2006).

Where to See It

Siberut macaques prefer riverine coastal swamp forest. They can be found in both primary and disturbed habitat. Siberut National park is the best place to see this animal. A German Primate Research Center in collaboration with Bogor Agriculture University built a research station there to study this species.

Access to Siberut:

- From Padang to Muara Siberut or to Muara Sikabaluan on Siberut Island takes about 10 hours by ferry boat (scheduled three times a week). To get from Muara Siberut/ Muara Sikabaluan to Siberut NP you can rent a motor boat to take you through the rivers for a leisurely trip of about 4-5 hours . There is also a faster boat, which takes approximately 4-5 hours.

In Siberut, there is accommodation at the Syahruddin Hotel and Wisma Tamu Guesthouse. Local guides are avalable in Padang, Bukittinggi and Muara Siberut. To find local guides you can ask at Jl. Pontianak N/13 Padang, West Sumatra.

23. Moor Macaque

Macaca maura

Other common names: Kera Hitam Dare; Lesang (Pinrang), Ceba (Bugis), Darre (Ujung Pandang).

Identification
The head and body length of *M. maura* is between 55 and 69 cm for males and between 45 and 59 cm for females. Their tails are short (males 5 to7 cm, females 2 to 4 cm). Body weight ranges from 8.2 to10 kg for males and 3.8 to 7.6 kg for females. Generally, the body color of this species is either brown or brown-black, which can be distinguished readily in the field. The ventral surface is often noticeably paler. They have a short muzzle with a dark-angular face with extremely prominent brow ridges and flat crown hair. The rump patch is not highly conspicuous; it is brownish-grey, considerably lighter in tone than the brown or black-brown of the rest of the body. This rump patch is not sharply set-off, not very broad, and not extending down the thighs to the knees (Groves, 2001, Mittermeier et al., 2013).

Geographic Range

The Moor macaque is only found in Sulawesi from Bontobahari at the southern tip of the southwest peninsula to north of Tempe Lake at Sakholi and Maroangin, South Sulawesi province.

Behavior and Ecology

The Moor macaque is frugivourous. They are known to feed on the fruit of Figs (*Ficus* sp) , Rao (*Dracontomelon mangiferum*), Pangi (*Pangium edule*), Bakan (*Litsea firma*), and many other fruit trees (Mittermeier et al., 2013). It also eats insects, fungi and some small mammals, in particular during the dry season. It is also recognized as an agricultural pest by local people.

Moor macaques jump from one tree to another and move quadrupedally when walking on branches or the ground. They mostly live on the ground because of the low tree density of the forests in which they live. Movement is initiated by the male group leader and followed by the other members of the group. They sleep on branches in their groups during the night. Moor Macaques live in multimale-multifemale groups consisting of between 15 and 40 individuals. Their home range is between 25 and 40 ha.

The Moor Macaque vocalization is unique among Sulawesian macaques. As they search for food, the group leader (adult male) makes a sound like a bird (pi .. pi .. pi ..). When another group is encountered, the sound becomes 'Ha' or 'Ga', as well as pi .. pi .. pi .. When this sound is heard, other group members will move toward to the source of sound or go quiet and show alert behavior. A barking sound will be made if they feel threatened (Supriatna et al., 1992).

Conservation Status

M. maura is currently thought to be the most threatened of Sulawesi's macaques because of the extent of habitat loss and human hunting. The IUCN/SSC Primate Specialist Group lists this species as Endangered (Eudey 1987). It was protected with the Indonesian government regulation by the Ministry of Forestry decree in 1991, No. 301/Kpts-II/1991.

Forest clearing has led to much habitat loss. Approximately 23,000 km^2 of original habitat has become only 2,800 km^2. Their population size is now less than 4,000 individuals (Supriatna et al., 1992, Supriatna et al., 2003). These macaques are also well known as raiders of crops such as corn, pineapple, papaya, and coconut, which brings them into conflict with villagers who will kill them (Supriatna et al., 1992).

Where to See It

Moor macaques live in primary or secondary forest up to 2000 m asl. They are also found in forest close to villages. The best place to see this macaque is in the Bantimurung Bulusaraung National Park, approximately 30 km from Makassar City in the direction of Chamba. It can be seen along the road in the forested areas between Maros and Chamba.

24. Celebes or Crested Black Macaque

Macaca nigra

Other common names: Yaki (Tonsea and Bacan), Wolai (Tondano), Bolai (Mangondow)

Identification

The Celebes Black macaque has a head and body length of between 50 and 57 cm for males and between 44.5 and 55 cm for females. They have a short tail of around 2 cm. The average weight of males is between 10.2 and 13 kg and of females between 5.5 and 8 kg (Mittermeier et al., 2013). They have narrow cheeks. Body coloration is generally dark brownish black covering the whole body. There is a distinct erect blackish crest on the crown between 5 and 15 cm in length. Arms, feet, and the ventral surface are darkish. Infants are born with a pink face that later will turn black. This species has a distinct pink kidney-shaped ischial callosity. Compared to other Sulawesi Macaques, *M. nigra* male scrota and anuses have red coloration.

Geographic Range

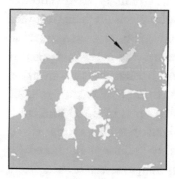

This species occurs in North Sulawesi, from Gunung Dua Sudara, Manembo-nembo, Kotamobagu and Mondayag in North Sulawesi Province. It also occurs on Bacan Island, where it was probably introduced from North Sulawesi. Primary rainforest is the best habitat for *M. nigra* though they also can be found in secondary forest and field gardens.

Behavior and Ecology

Like other macaques, this species eats various parts of plants such as leaves, flowers, seeds, and fruit. In Tangkoko Batuangus Nature Reserve, a group of macaques is often seen searching for mollusks. This macaque also eats crops such as coconut, mango, sugar palm sap, and others (Mittermeier et al., 2013). They move mainly on the ground, spending around 60% of their time on the ground for travel or to rest. Movement is mostly quadrupedal, whether on the ground or branches. This macaque is active during the day. Its home range is between 74 and 350 ha, depending on the group size and time of the year. Their social organization is multimale-multifemale. Group size is between 60 and 80 individuals, which is quite large compared to other macaques in Sulawesi with an adult ratio of males to females of 1: 3. Their vocalization is a 'Ko Ko Ko Ko' sound. As with other primates generally, sound is used as sign of danger or for showing power to members of other groups.

Conservation Status

This macaque is protected by the government of Indonesia with the Ministry of Agricultural decision letter 29 January 1970 No 421/Kpts/um/8/1970, the Ministry of Forestry decision letter 10 June 1991 No 301/Kpts-II/1991 and UU No. 5 1990. It is categorized as 'Critically Endangered' on the IUCN Redlist and listed in Appendix II of CITES. It has lost 60% of its original habitat due to conversion of forest to agriculture and plantations, from 12,000 km² to 4,800 km². It also occurs in several conservation areas of 2,750 km2 such as Menembo-nembo and Tangkoko Batuangus (Supriatna et al., 2003). The population is declining sharply because it is hunted for its meat. In Minahasa, especially during religious ceremonies, local people consume many species of animal including this macaque. This species is also threatened by the illegal pet trade. Southwick et al., (1994) estimated that in 1991 the population was no more than 2,000 individuals.

Where to See It

A healthy population of this macaque can be found in Tangkoko Batuangus Nature Reserve and also probably in Sungai Onggak Dumoga where their distribution likely becomes adjacent to that of *Macaca nigrescens*. They are found in primary or secondary lowland forest mainly in coastal areas where they can forage for favorite fruit, e.g. *Dracontomelon* spp. and *Ficus* spp. But they are also found up to 2,000 m above sea level. They often visit village gardens for food and can destroy the harvest, which is why they are regarded as agricultural pests. The easiest place to reach for finding Yaki is Tangkoko Batuangus Nature Reserve and Dua Saudara Natura Reserve which is 60 km from Manado, North Sulawesi. It can also be seen at Gunung Lokon Nature Reserve, Gunung Ambang and Tangale, approximately 100 km to the west of Manado.

25. Gorontalo Macaque

Macaca nigrescens

Other common name: Dihe (Gorontalo)

Identification

The Gorontalo Macaque has a head and body length of around 60 cm for males and around 50 cm for females. Both have a short tail of around 2.5 cm. They have a baboon-like face. They have a narrow erect elongated crest that is lacking in infants but will grow with age up to 5 to 10 cm.

This is slightly shorter that of *M. nigra*. The limbs of adult Gorontalo macaques are blackish, which contrasts with the overall brown body coloration. In the young, limbs are pale brown but they will darken over time.. Callosities are an elongated oval shape of a gray color without any internal subdivision.

Geographic Range

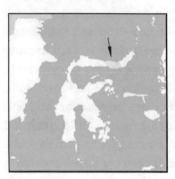

Broadly, the range of this species is the middle part of the northern Sulawesi peninsula, from the boundary of Gorontalo Province in the west to Sungai Onggak Dumoga in the east. They are usually found in lowland to highland and mountainous areas.

Behavior and Ecology

The Gorontalo macaque is known to eat approximately 69 different kinds of fruit (Kohlass, 1993). About 70% of its diet consists of fruit, and the rest consists of other plant parts, insects, mollusks and other small invertebrates. Like other species of Macaque, the Gorontalo macaque has a special pouch in the cheek for storing food.

They live in groups of approximately 20 individuals. Sometimes there is a smaller group of 2 to 5 individuals, but there are also large groups of 60 individuals. The male to female ratio in a group is 1:2. They spend a lot of time in trees, but when travelling they come to the forest floor, especially when moving fast. Movement is mostly quadrupedal. There are four movement types, jumping, crawling, vertical hanging, and sometimes walking bipedally.

Gorontalo macaques are active during the day or diurnally. The day activity mostly occurs on the ground, and they sleep at night in trees. In areas surrounding Bogani Nani Wartabone National Park, they often raid plantations, especially corn. This results in reducing the population, either by spreading poisoned corn or hunting.

The Gorontalo macaque produces a sound that is similar to that of *M. hecki,* which is one syllable (pi..) not repeated, on and off. Like other macaques, the sound acts as a communication tool among the members of groups. When hearing their voices, the villagers are alert to save their plantations from raids (Kohlaas, 1993).

Conservation Status

This species is recorded as Vulnerable on the IUCN Red list. It is listed in Appendix II of CITES. The area of available habitat has shrunk by about 60% from more than 12,000km^2 to only around 4,800 km^2 (Supriatna, 2015). Due to a rapidly increasing indigenous population, together with an influx of transmigrants and an increase in logging operations, the area of forests has steadily declined. This species is hunted for bushmeat and culled as pests. This primate, endemic to Sulawesi, has been protected by the government of Indonesia, based on the Ministry of Agriculture decision letter 29 January 1970 No 421/Kpts/um/8/1970, the Ministry of Forestry decision letter 10 June 1991 No 301/Kpts-II/1991 and UU No. 5 1990.

Where to See It

Gorontalo macaques generally live in the lowland and hilly forest, between altitudes of 400 to 600 m above sea level. Nani Bogani Wartabone National Park, Kotamubagu, North Sulawesi, is the easiest place to see this primate. But along the road from Kotamubagu toward Gorontalo, you may find macaques on the roads in the forested areas. Nani Bogani Wartabone NP is easy to get to. From Manado or Gorontalo go in the direction of Kabupaten Kotamubagu. From Kabupaten Kotamubagu, you can take a rental car to Torout where the Nani Bogani Wartabone National Park office is located. Some small hotels are available surrounding the national park. This destination is also very famous for bird watching.

26. Heck' Macaques

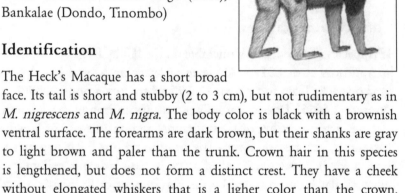

Macaca hecki

Other common names: Dige (Buol), Bankalae (Dondo, Tinombo)

Identification

The Heck's Macaque has a short broad face. Its tail is short and stubby (2 to 3 cm), but not rudimentary as in *M. nigrescens* and *M. nigra*. The body color is black with a brownish ventral surface. The forearms are dark brown, but their shanks are gray to light brown and paler than the trunk. Crown hair in this species is lengthened, but does not form a distinct crest. They have a cheek without elongated whiskers that is a ligher color than the crown. Heck's Macaques have ischial callosities but differ from other Sulawesi macaques, in that they are gray to yellow in color. Ischial callosities form kidney-shapes, with only a partial transverse furrow across them. The males head and body length is between 58 and 68 cm and they weigh between 8 and 10 kg. The females body length is between 50 to 57 cm and weight between 7 and 8 kg.

Geographic Range

Heck's Macaques are distributed broadly in northern peninsula Sulawesi. They have a southern boundary at Kampung baru and an eastern boundary at Kwandang.

Behavior and Ecology

The diet of Heck's macaque is almost the same as other Sulawesian macaques, which is many parts of plants, such as leaves, sprouts, flowers, fruit and tubers, and some species of insects, some mollusks, and even small vertebrates. They live within multimale-multifemale groups comprising 10 to 15 individuals. Males will leave their natal group while females tend to stay in the group.

This species has a large home range of more than 100 ha. They can be arboreal, but they also explore the forest floor using all four limbs to move, i.e. quadrupedally. Heck's Macaque moves faster on the forest floor than in trees. Trees are used for moving, eating and sleeping. The sleeping place is usually on the main branch of a tree.

Generally, they are active during the day. They often visit the gardens and crops of villagers so like other Sulawesian macaques, are considered to be pests. At Panua Nature reserve, Heck's macaques invade corn plantations and are therefore hunted by villagers. Before entering the plantation, the head male usually observes from a tree, and other group members stay on the ground. After entering the plantation, the dominant male will come to the ground and eat but remain in an alert manner (pers observation).

The vocalization of Heck's Macaque is a long repeated screaming sound (pi ... pi ... pi) louder even than *M. maura*. The sound functions as an alert to danger or is used by males in particular to confront other groups (pers observation).

Conservation Status

Like other Sulawesian macaques, Heck's macaque suffers from habitat loss. More than 33% of the 67,000km^2 of their habitat has been lost. Habitat loss, along with increasing human pressure resulting from shifting agriculture, an influx of migrants from other parts of Indonesia and logging operations, are the main factors threatening this species.

At present, there are only several small conservation areas up to 1,055 km² to help with their protection. They are listed as vulnerable on the IUCN, Red list and they are listed under Appendix II in CITES. They are also protected by the Ministry of Agricultural decision letter 29 January 1970 No. 421/Kpts/um/8/1970, the Ministry of Forestry decision letter 10 June 1991 No. 301/Kpts-II/1991, and UU No. 5 1990.

Where to See It

Heck's macaque lives in the lowland tropical forest up to between 1600 and 1800 m above sea level. This macaque can be seen at the Bogani Nani Wartabone National park about 50km west of Kotamobagu. It can also be seen at Paguyaman river, at Nantu Forest and in forested areas along the Trans-Sulawesi road between Gorontalo and Palu (Central Sulawesi) such as the Panua protected area.

27. Tonkean Macaques

Macaca tonkeana

Other common names: Boti (Poso), Seba (Tana Toraja), Lesang (Pinrang)

Identification

According to Fooden (1969) there are 2 subspecies of *Macaca tonkeana*: *M. t.* *tonkeana* and *M. t. togeanus* (occurring only in the Togean islands). Groves (2001) mentioned that this species hybridizes with *M. maura* and *M. hecki* where their ranges overlap. In 1999, Froehlich and Supriatna (1996) named the population in the Balantak mountain area, in eastern Central Sulawesi, *Macaca balantakensis*. However, this is almost certainly a hybrid swarm, rather than a distinct species.

The tonkean macaque has a body length of between 50 and 70 cm with tail length between 3 and 7 cm. Their body weight ranges between 12 and 14 kg. Dorsal body coloration is shining black. The head is brown to dark brown with little crown hair is also dark brown. This species has the blackest hair of all the macaques in Sulawesi. The leg hair is short with the ventral part dark brown to black. The young have darker coloration on the neck and head compared to the adults. In this species there is no difference in color betweenmale and female. The Tonkeana macaque has an oval-shaped pink ischial collosity but there is no separate ridge.

Geographic Range

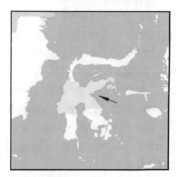 The distribution of the Tonkean macaque is quite wide compared to other macaques from Sulawesi. This species can be found in Central Sulawesi, from the lowlands of Siweli-Kasimbar in the north to Tempe lake in the south-west, and Matana and Towuti lakes in the south-east. This species has been seen up to 1,300 m above sea level in the Fehrumpenai Nature Reserve.

Behavior and Ecology

Tonkean macaques eat plant parts, especially fruit (56,77% of their diet) and leaves (16,7 %) (Supriatna et al., 1992). They also consume plant sprouts, grass and many species of insect, mollusk and small vertebrates. In the wild, they live in groups that can reach up to 25 to 40 individuals. Although they form social systems, there is no hierarchy among the members.

During daily activities, this primate is often seen on the ground and moves using all four limbs. They are active foragers during the

The tonkean macaque has a body length of between 50 and 70 cm with tail length between 3 and 7 cm. Their body weight ranges between 12 and 14 kg. Dorsal body coloration is shining black. The head is brown to dark brown with little crown hair is also dark brown. This species has the blackest hair of all the macaques in Sulawesi. The leg hair is short with the ventral part dark brown to black. The young have darker coloration on the neck and head compared to the adults. In this species there is no difference in color betweenmale and female. The Tonkeana macaque has an oval-shaped pink ischial collosity but there is no separate ridge.

Geographic Range

The distribution of the Tonkean macaque is quite wide compared to other macaques from Sulawesi. This species can be found in Central Sulawesi, from the lowlands of Siweli-Kasimbar in the north to Tempe lake in the south-west, and Matana and Towuti lakes in the south-east. This species has been seen up to 1,300 m above sea level in the Fehrumpenai Nature Reserve.

Behavior and Ecology

Tonkean macaques eat plant parts, especially fruit (56,77% of their diet) and leaves (16,7 %) (Supriatna et al., 1992). They also consume plant sprouts, grass and many species of insect, mollusk and small vertebrates. In the wild, they live in groups that can reach up to 25 to 40 individuals. Although they form social systems, there is no hierarchy among the members.

During daily activities, this primate is often seen on the ground and moves using all four limbs. They are active foragers during the morning until afternoon, but sometimes around noon they spend time playing. Movement from one tree to another is usually by jumping. The whole group moves simultaneously and when they do, can be very noisy. Daily movement can reach up to 1,100 m. Only the adult males produce a loud calling sound, which is almost similar to *M. maura*, but more often and repeated.

As their natural habitat is declining, this species is known to enter plantations on the periphery of the forest. Therefore, they are considered to be pests. In coconut plantations, they are known for their ability to peel young coconuts fast. They usually only eat the coconut meat and drop the rest, but sometimes they take the coconut into the forest. They show considerable flexibility in response to anthropogenic disturbance by adjusting their use of forest strata to facilitate travel and increase foraging opportunities and by intensively using particular areas within their home range where known resources are present and predictably available (Riley, 2008).

Loud calls in Tonkean macaques function primarily to influence intragroup cohesion; that is, to communicate information with regard to location and movement within the social group. The between-group differences observed may reflect differences in group size, and social and ecological conditions (Riley, 2005).

Conservation Status

The Tonkean macaque is listed as vulnerable on the IUCN Red list. Their population is more secure than other macaques of Sulawesi due to their wide and extensive distribution. This species is well represented within protected areas such as Lore Lindu National Park, and Fehrumpenai Nature Reserve with good forest, totaling more than 550,000 ha. The Tonkeana macaque is protected by the Ministry of Agricultural 29 January 1970 No. 421/Kpts/um/8/1970, the Ministry of Forestry 10 June 1991 No. 301/Kpts-

Behavior and Ecology

The diet of Heck's macaque is almost the same as other Sulawesian macaques, which is many parts of plants, such as leaves, sprouts, flowers, fruit and tubers, and some species of insects, some mollusks, and even small vertebrates. They live within multimale-multifemale groups comprising 10 to 15 individuals. Males will leave their natal group while females tend to stay in the group.

This species has a large home range of more than 100 ha. They can be arboreal, but they also explore the forest floor using all four limbs to move, i.e. quadrupedally. Heck's Macaque moves faster on the forest floor than in trees. Trees are used for moving, eating and sleeping. The sleeping place is usually on the main branch of a tree.

Generally, they are active during the day. They often visit the gardens and crops of villagers so like other Sulawesian macaques, are considered to be pests. At Panua Nature reserve, Heck's macaques invade corn plantations and are therefore hunted by villagers. Before entering the plantation, the head male usually observes from a tree, and other group members stay on the ground. After entering the plantation, the dominant male will come to the ground and eat but remain in an alert manner (pers observation).

The vocalization of Heck's Macaque is a long repeated screaming sound (pi ... pi ... pi) louder even than *M. maura*. The sound functions as an alert to danger or is used by males in particular to confront other groups (pers observation).

Conservation Status

Like other Sulawesian macaques, Heck's macaque suffers from habitat loss. More than 33% of the 67,000km^2 of their habitat has been lost. Habitat loss, along with increasing human pressure resulting from shifting agriculture, an influx of migrants from other parts of Indonesia and logging operations, are the main factors threatening this species.

At present, there are only several small conservation areas up to 1,055 km² to help with their protection. They are listed as vulnerable on the IUCN, Red list and they are listed under Appendix II in CITES. They are also protected by the Ministry of Agricultural decision letter 29 January 1970 No. 421/Kpts/um/8/1970, the Ministry of Forestry decision letter 10 June 1991 No. 301/Kpts-II/1991, and UU No. 5 1990.

Where to See It

Heck's macaque lives in the lowland tropical forest up to between 1600 and 1800 m above sea level. This macaque can be seen at the Bogani Nani Wartabone National park about 50km west of Kotamobagu. It can also be seen at Paguyaman river, at Nantu Forest and in forested areas along the Trans-Sulawesi road between Gorontalo and Palu (Central Sulawesi) such as the Panua protected area.

27. Tonkean Macaques

Macaca tonkeana

Other common names: Boti (Poso), Seba (Tana Toraja), Lesang (Pinrang)

Identification

According to Fooden (1969) there are 2 subspecies of *Macaca tonkeana*: *M. t. tonkeana* and *M. t. togeanus* (occurring only in the Togean islands). Groves (2001) mentioned that this species hybridizes with *M. maura* and *M. hecki* where their ranges overlap. In 1999, Froehlich and Supriatna (1996) named the population in the Balantak mountain area, in eastern Central Sulawesi, *Macaca balantakensis*. However, this is almost certainly a hybrid swarm, rather than a distinct species.

Where to See It

The Tonkean macaque lives in lowland primary forest, secondary forest and highland forest up to 1,300 m above sea level, such as in the Fehrumpenai nature reserve area. They are easier to observe in Lore Lindu National Park, about 70 km south of Palu, Central Sulawesi. They can also be found in agricultural areas, plantations, and coastal areas. Sometimes they can be seen along the Trans Sulawesi road where it runs through Fehrumpenai Nature Reserve to the South of Palu and in coffee plantation areas, about 40 km north of Palu.

28. Booted Macaques

Macaca ochreata

Other common names: Hada (Tolaki), Ndoke (Buton).

Identification

The head and body length of males is between 48 and 59 cm and females is between 40 and 48 cm with a male tail length of 4 to 6 cm and a female tail length from 3 to 5 cm (Mittermeier et al., 2013). Body color is blackish or brown above with forearms, shanks and rump whitish or light brown. They have darker hair on the crown that stands erect and forms a rectangular cap. They have oval ischial callosities that often bend outward. There are two subspecies with slightly different coloration (Corbet & Hill, 1992):

- *M. o. ochreata*
  This subspecies has a generally blackish body color and is darker than *M. o. brunnescens*. Forearms, shanks and rump in this subspecies are rather whitish in color.
- *M. o. brunnescens*
  M. o. brunnescens has a lighter generally brown body compared to *M. o. ochreata*. Forearms, shanks and rump are lighter brown

in color. This subspecies also has a shorter face. This subspecies is also known as the Buton macaque.

Geographic Range

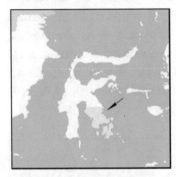

- *M. o. ochreata*

This subspecies ranges in the mainland of the entire South East of Sulawesi

- *M. o. brunnescens*

This subspecies is found on Muna Island, Butung (also known as Buton Island) and possibly on the neighboring islet of Pulau Labuan Blanda, off the south-west coast of Sulawesi. It may no longer occur on Muna as the island has been completely deforested.

Behavior and Ecology

From a short-term study (Bismark, 1982) , this species was observed to consume fruit (60%), stems and flowers (20%), leaves (12%) and invertebrates (2%). Like other primates in Sulawesi, this species is considered to be an agricultural pest, especially of the farms on the periphery of the forest. They also often raid village crops such as cacao, sweet potato, bananas and corn. Groups of this species who raid crops spend more time grooming and resting than groups in the forest, which spend most of their time travelling.

They are active during the day and spend about the same amount of time in trees as on the ground. Movement on the ground is quadrupedal. Their also depends on tree density. When in dense forest, they rarely move on the ground. When on secondary forest, where tree density is lower, they often come to the ground exploring. During exploring, they are silent, especially when entering the village agricultural areas. The alpha male is always alert to keep watch on the group. The sound is louder when danger threatens, such as from dogs and humans.

They are organized socially into multimale-multifemale groups, which range in size from 12 to 28 individuals for the nominate species and 12 to 23 individuals for *M. o. brunnescens*. The mean home range size of this species is around 62 hectares.

Conservation Status

M. ochreata, including both subspecies *M. Ochreata ochreata* and *M. ochreata brunescens*, is listed as Vulnerable on the IUCN Redlist and protected in CITES Appendix II. Both subspecies are threatened by habitat conversion for human uses. The Buton macaque has lost more than 37% of its habitat from 29,500 km² to 18,500 km². Like the other macaques of Sulawesi, this species has also been protected by the Ministry of Agricultural decision letter 29 January 1975, No. 35/Kpts/um/1975 and the Ministry of Forestry 10 June 1991 no. 301/Kpts-II/1991, and UU no 5 1990. There is a conservation area, Rawa Aopa National Park, of 1,429 km2, which offers some protection for this species.

Where to See It

The booted macaque occupies lowland tropical forest to an altitude of between 600 and 800 m above sea level. It can be found around plantations and live on the periphery of agricultural areas. The subspecies Buton macaque lives in lowland primary and secondary forest from 0 to 200 m above sea level. They are often seen on the periphery of plantations or agricultural areas.

Booted macaques can be found in Rawa Aopa National Park that can be reached from Kendari, Southeast Sulawesi in 2 to 3 hours. A direct flight from Jakarta to Kendari is now available. But the flight to Kendari through Makassar is also still available. From Kendari you can rent a car to reach Rawa Aopa National Park.

Figure 6. Indonesian Presbytis: a. *Presbytis femoralis* (Suprayitno); b. *Presbytis hosei* (TN Kayan Mentarang & Nanang K. Hadi); c. *Presbytis natunae* (Ferdi Rangkuti); d. *Presbytis siamensis* (Badrul Munir & Tatang Mitra Setia); e. *Presbytis comata* & *Presbytis comata fredericae* (Anton Ario & Tatang Mitra Setia); f. *Presbytis potenziani* (Reki Kardiman); g. *Presbytis siberu* (Gusmardi Indra); h. *Presbytis mitrata* (Nurul L. Winarni)

Figure 7. Indonesian Presbytis continued: i. *Presbytis bicolor* (Sri Mariati & Arif Rudianto); j. *Presbytis frontata* (Indra Hana & Usman); k. *Presbytis chrysomelas* (Asri Ali & M. Rizki Gumelar); l. *Presbytis rubicunda* (Susan Cheyne, Ehler Smith, & Indra Hana); m. *Presbytis thomasi* (Jatna Supriatna)

Figure 8. Indonesian Presbytis: n. *Presbytis canicrus* (Nur Rohman); o. *Presbytis melalophos* (Ardika Dani Irawan); p. *Simias concolor* endemic Mentawai Islands (Gusmardi Indra); q. *Nasalis larvatus* (Sofian Iskandar), endemic Kalimantan; r. *Trachypithecus auratus* (Sri Mariati); s. Trachypithecus mauritius (Anton Ario); t. Trachypithecus cristatus (Sri Suci Utami Atmoko & Indra Hana)

2. Subfamily: Colobinae

The subfamily Colobinae has 10 genera, but only four of those are found in Indonesia, including the leaf monkeys (*Presbytis* and *Trachypithecus*) and the odd-nosed group (*Nasalis* and *Simias*).

Leaf monkey (Langur) group: *Presbytis* and *Trachypithecus*

As the name indicates, a large part of the leaf monkey's diet consists of leaves, accompanied by fruit, seeds and some insects. They have a sacculated stomach, consisting of three or four chambers which enables the breakdown of cellulose through fermentation. Morphologically, the genus *Presbytis* is characterized by its underbite incisor occlusion. Its incisors are narrow, with sharp and high crest molars (Groves 2001). The incisors of *Presbytis* are relatively large as they consume more seed than the closely related genus *Trachypithecus*.

Odd-nosed group: *Nasalis* & *Simias*

The genera *Nasalis* and *Simias* are endemic to different islands in Indonesia. *Nasalis* is only found on Kalimantan while *Simias* is only found on the Mentawai Islands. Although they are separated by 1,000 km, they are closely related. Both genera are monotypic, which means there is only one species within the genus. *Nasalis* males have a distinctive large, pointed nose and females have a tapered nose. Male body size is larger than that of females. The *Simias* have short noses that point upwards. They have short pig-like tails and are also called pig-tailed langurs.

29. Javan Langur

Presbytis comata

Other common names: Surili for *P. c. comata* (West Java); and Rekrekan for *P. c. fredericae* (Central Java)

Identification

Taxonomists have divided this species into two subspecies: *Presbytis comata comata* and *Presbytis comata fredericae*, based on differences in coloration and size, although appearances have been described as similar. Previously, Brandon Jones et al., (2004) identified these two subspecies as distinct species, but a recent publication by Ross et al., 2013 recognized both as subspecies of *P. comata*.

- *P. c. comata* (Javan Langur, sometimes called Javan Grizzled Langur) In general, adult body coloration in the head and centro-dorsal region is black, brown, and greyish. They have a distinct short grizzled hair patch on their rumps. In infants, coloration is usually white with a dark stripe from head to tail tip. As the langur ages, the dark area gradually extends until only the underparts remain white; or the inner thigh remains white, with a white strip continuing down the leg to the ankle. They have prominent nasal bones and a variety of crests on the crown of the head have been observed. The average body mass of an adult male is around 6.7 kg and of female around 6.4 kg (Fleagle,1988). Both subspecies have enlarged salivary glands, and a sacculated stomach to help break down cellulose.

- *P. c. fredericae* (Javan Fuscous Langur) *P. c. fredericae* have a black back; with white or light grey throat, upper chest, abdomen, arms, inside of legs, extending to the tail (Brandon-Jones, 1995; Nijman, 1997). On the thumbs, middle fingers and sometimes distal phalanges of the digits there is a small amount of white (Brandon-

Jones, 1995). The pelage coloration of infants varies from dark grey to black (Nijman, 1997).

Geographic Range

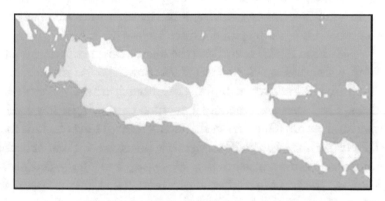

• *P. c. comata (Javan Grizzled Langur)* This subspecies is found in West Java (Nijman 1997).

• *P. c. fredericae (Javan Fuscous Langur)* This subspecies is found in Central Java (Nijman, 1997) It is only known from Mt. Slamet, the northwestern slopes of the mountains north of the Dieng plateau, and Mt. Lawu (Nijman and Sozer, 1995). In the Dieng mountains, this subspecies is found to live in primary and secondary forests, at the edges and the interior, and in lowland forests, forests on steep slopes and hills, and upper montane forests (Nijman and van Balen 1998). In the Dieng mountains this subspecies has been observed at an altitude of 2,565 m (Nijman and Sozer, 1995).

Behavior and Ecology

Javan langurs of both subspecies are normally diurnal, meaning they are most active during the day with three to four feeding peaks per individual per day (Ruhiyat, 1983).. They are semi-brachiators, moving through trees by swinging from branch to branch, with the capacity to make spectacular leaps. Brachiation is a specialized form of arboreal locomotion, used by primates to move very rapidly while hanging beneath branches. Arguably the epitome of arboreal locomotion, it involves swinging with the arms from one handhold to another. Only a few species are brachiators

Ruhiyat (1983) found that the diet of this species consisted of 59.1% young leaves, 13.5% fruit, 7.0% flowers, 5.6% mature leaves, 4.1% fungi, 2.7% pseudo bulbs, 1.5% branch tips, 0.7% seeds and others. The leaves are generally immature, containing a low level of lignin and tannin (Gurmaya et al., 1994). Preferred leaves include those of *Ficus pubinervis, Passiflora ligularis, Elaegnus triflora, Schefflera aromatica, Jasminum azoricum, Hoya sp.*, and *Aeshynanthus sp.* (Ruhiyat, 1983). Leaves of epiphytes and lianas in the lower and middle layers of the forest and flowers of *Pandanus furcatus* and petioles of *Alsophylla glauca* are also eaten (Ruhiyat, 1983).

During the fruiting season this species is attracted to orchards and other solitary fruiting trees (Melisch and Dirgayusa, 1996). Preferred fruits include: *Premna parasitica, Pygeum* spp., *Saurauia* spp., and *Castanopsis argentea* (Ruhiyat, 1983).

This species rarely drinks as it receives most of its water from the food it eats. Javan langurs have been observed to come to the ground and eat reddish-colored soil (Ruhiyat, 1983). This behavior is called geophagy. Many animals eat some soil to help balance their diets.

The social structure of the Javan langur is variable and appears to be affected by disturbance. In non-disturbed areas, groups are polygamous, while in disturbed areas, they are primarily monogamous. They tend to live in small groups and the few intergroup encounters that occur are generally aggressive. Adult males are usually the most aggressive and are frequent participants in these intergroup encounters. Groups have small overlapping home ranges.

Group sizes for this species range from 3 to 20 individuals. At Gunung Tukung of Gunung Gede Pangrango National Park, the size of groups ranged from 5 to 23 individuals (Melisch and Dirgayusa, 1996). In Central Java, Nijman (1997) found that group sizes on Gunung Slamet ranged from 4 to 10 individuals; while at Mt. Prahu, group sizes for this species ranged from 2 to 13 individuals (Nijman and van Balen, 1998). In Kuningan forest, it was found that the group size varied from 2 to 22 with an average of 8.52 animals. The population density at the confidence interval of 95% ranged from 44.39 to 82.36 animal/km^2 (average was 60.47) (Supartono et al., 2016).

Javan langurs have a unimale social system and a polygynous mating system. More than one male has been observed in a group even though there is usually only one (Ruhiyat, 1983). This is a territorial species that has aggressive encounters with conspecific groups. A group's home range will overlap with the home ranges of other groups, and intergroup aggression occurs in overlap zones when preferred foods are present. The adult male is dominant over all other members of the group (Ruhiyat, 1983). Females perform most of the grooming bouts in the group. Males disperse from the natal group before adolescence. This species has been found to form mixed-groups with the ebony leaf-monkey, *Trachypithecus auratus* (Nijman, 1997).

Vocalization in this species is a fast sequence of 20 "kik" notes heard within 2.5 seconds. It is emitted by adult males during intergroup encounters (Ruhiyat, 1983). An adult female would emit this call when she missed her infant, but the call is weaker and shorter (Ruhiyat, 1983). When the adult males utter this call, other group members move to the upper levels of the canopy (Ruhiyat, 1983). Leopards (*Panthera pardus*), and fishing cats (*Prionailurus viverrinus*)are assumed to be their predators (Melisch and Dirgayusa, 1996; Seidensticker, 1983; Ruhiyat, 1983).

Conservation Status

Both *P.c. comata* and *P. c. fredericae* are listed as Endangered (under *Presbytis comata*) on the IUCN Redlist. They are also listed in CITES Appendix II. This Javan Langur endemic is protected by law by the Indonesian government, and is classified as endangered, due to the dramatic habitat loss it has suffered (Supriatna et al., 1994). The human population growth rate in West Java province is among the highest in Indonesia and remaining forest is scarce. The species is protected in several National Parks in West Java (Ujung Kulon, Gunung Halimun, and Gunung Gede-Pangrango, Ciremai) and some protected areas in West and Central Java.

Where to See It

This species can be found in primary and secondary forests, ecotones, in the forest interior, lowland forests, forests on steep slopes and hills, and in montane and upper montane forests. It also can be found in plantations and orchards (Melisch and Dirgayusa, 1996; Seitre and Seitre, 1990). At Gunung Pongkor, Mt. Halimun National Park, this species was found to occur in hill forests from 500 to 1000 m in altitude (Indrawan et al., 1995).

Javan langurs can be seen in conservation areas such as Gunung Gede Pangrango National Park, Gunung Halimun Salak NP, Ujung Kulon NP, Situ Patenggang Protected Area, Kawah Kamojang Protected Area, and Ranca Danau in Banten. Gunung Gede Pangrango NP is an easy place to observe this primate. Its only 2 hours drive from Jakarta. The two best places to see it are Cibodas, near the national park office and the Bodogol Conservation Education Center. To reach this education center, you can go to Sukabumi from Bogor (15 km), which is behind the Lido Hotel. In this education center you can also observe javan gibbons, long-tailed macaques and javan slow lorises.

This area has a 5 km hiking trail that makes it easy to look for these primates.

30. Pale-thighed Langur

Presbytis siamensis

Other common names: Kokah (Riau), Kera cantik and Kekah (Bintan Island)

Identification

Presbytis siamensis has four subspecies. Three of them, *P. s. rhionis*, *P. s. cana*, and *P. s. paenulata* are found in Indonesia. The fourth, *P. s. siamensis* occurs in Peninsular Malaysia and South Thailand. In general, *Presbytis siamensis* are pale grayish brown in color and have blackish hands, feet and brow. They have a whitish underside, which continues with a grayish-whitish coloration onto the outer side of the thigh. They have a white whorl on the forehead. The scrotum of the male is black (Curtin, 1980). The infant has white coloration with a cruciform blackish pattern on the dorsal side (Wilson and Wilson, 1976; Aimi and Bakar, 1996).

Their incisors are narrow and molars are sharp. The dental formula for this species is 2:1:2:3 on both upper and lower jaws (Ankel-Simons, 2000). This formula means 4 incisors, 2 canines, 4 premolars and 6 molars in each jaw. The jaws are deep but they have a short and broad face (Oates and Davies, 1994). This species has a recluded thumb with widely spaced orbits and hind limbs that are longer than the forelimbs (Davies, 1991). The average adult body mass is 6.0 kilograms (Oates and Davies, 1994).

The subspecies that are found in Indonesia have different pelage colorations:

- *Presbytis siamensis rhionis* (Bintan White-thighed surili). The pelage coloration of this subspecies is brown and gray (Wilson and Wilson, 1976). P. s. rhionis (Miller) is found on Bintan island in the Riau Islands.

- *Presbytis siamensis paenulata* (Mantled pale-thighed surili). This subspecies has a pelage color that is dark brown on the dorsal side and pale-gray brown on the ventral side and inner limbs (Wilson and Wilson, 1976). The limbs are brown except for white on the outer thighs (Wilson and Wilson, 1976). The hands and feet are black (Aimi and Bakar, 1992). Surrounding the eyes there are white eye-rings and on the top of the head there is a reduced crest (Wilson and Wilson, 1976).
- *Presbytis siamensis cana* (Riau pale-thighed surili). The pelage color of this subspecies is brown and gray (Wilson and Wilson, 1976).

Geographic Range

This species occurs in Indonesia (eastern Sumatra and the Riau Archipelago in the Strait of Malacca), Peninsular Malaysia, and extreme southern peninsular Thailand. Population isolates are known within the range of *P. femoralis* in Thailand and *P. melalophos* in Sumatra (Groves, 2001). The three Indonesian sub-species, all from Sumatra, are listed below:

- *Presbytis siamensis rhionis*
  Known for certain only from Bintan island, in the Riau Archipelago, Indonesia; it might occur on Batam island and Galang island as well (Groves, 2001).
- *Presbytis siamensis cana*
  Found in eastern Sumatra between the Siak and Indragiri Rivers, and on Palau Kundur in the Riau Archipelago (Groves, 2001).

- *Presbytis siamensis paenulata*

Found in east-central Sumatra, where it is confined to a small wedge of coastal forest, with a population isolate reported from near Lake Toba (Groves, 2001).

Behavior and Ecology

This species is primarily furgivorous but the diet includes leaves and flowers as well as fruit. Seeds are the part of the fruit most often consumed. The leaves consumed are mostly young leaves. Groups forage together as a cohesive unit. During foraging they sometimes break up into smaller sub-groups. Feeding time is spread throughout the day. There are two major peaks of feeding behavior, in the early morning and late afternoon. Fruits are consumed more in the early morning and early afternoon and new leaves are consumed more in the late afternoon.

Presbytis siamensis spends most of its time resting. A group tends to sleep and rest together. This species will use sleeping sites more than once. Sleeping sites also tend to be in trees bordering rivers.

Ranging behavior in this species consists of travelling in irregular loops in and out a central core area. These loops seem to be wider in the fruiting season (June – August) and shorter in the dry season (January – March). They move through the tree canopy during the day.

Group sizes vary from 2 to 8 individuals. Intergroup encounters increase when food sources are limited. Adult males tend to spend more time near adult females than do juveniles. Lone males tend to be restricted to less desirable parts of the forest. When habitat is disturbed by logging, this species will shift their home range, use the lower to middle canopy, and increase the level of folivory.

Conservation Status

The sub-species found in Indonesia are listed as Near Threatened on the IUCN Redlist, except for *P. s. rhionis* which is listed as Data Deficient due to the lack of information on its taxonomy and distribution. Extensive habitat loss has taken place within the range of the species but there is no evidence that their population is declining. However, on Bintan, locals report that the sub-species is rarely seen in the wild due to habitat loss. Most forests on Bintan Island has been converted into housing and agricultural land, so only a few hectares remain. Reassessment of this species is necessary once its taxonomy has become clearer.

Where to See It

They can be seen in primary and secondary forests, swamp forest, mangrove forest, and rubber tree plantations. Tesso Nilo National Park in Riau is a great place to see this primate. You can see them in local plantations close to the national park. To reach this national park, you have to use a four-wheel drive vehicle as the road is quite challenging. You can fly to Pekanbaru, and from there it can take between 4 and 5 hours through Tembilahan and then through an Asia Pulp Paper (APP) company concession, for which a special permit letter is needed. Without the permit letter, you can take a road around the APP concession, which will take a few hours longer depending on the road condition.

31. Banded Langur

Presbytis femoralis

Other common names: Kokah (Riau), Nokah (Northern Sumatra)

Identification

There are three sub-species of *Presbytis femoralis*, only one of which, the East Sumatran Banded Langur, *P. f. percura*, is found in Indonesia, Of the other two, *P. f. femoralis* is found in Peninsular Malaysia and Singapore, and *P. f. robinsoni* is found in Myanmar, Thailand, and Malaysia. Sub-species vary primarily with respect to pelage coloration, ranging from pale to dark. Generally, this leaf monkey has lighter dark-colored dorsal and pale beige-colored ventral sides. The beige coloring spreads from the inner thighs around to the posterior surface, hence restricting the dark area to the outer surface.

The sub-species found in Indonesia, the East Sumatran Banded Langur, is black on both the dorsal and ventral sides with the exception of a thin white ventral stripe (Wilson and Wilson, 1976). This thin stripe extends to cheeks, chin, wrists, and ankles but not onto the outer surface of limbs and is indistinct under the tail. Their chest is black with hair directed backwards. The head, face and muzzle are gray with narrow pale eye-rings. There is one or occasionally a pair of whorls (Groves, 2001).

Geographic Range

It occurs in a small area between the Rokan and Siak Rivers in Riau Province in the eastern part of the island of Sumatra.

Behavior and Ecology

Banded Langurs are diurnal, being active during the day. They move arboreally and are semi-brachiators in the tree canopy. They may also be seen on the ground, but this is rare. They have two daily feeding peaks, one in early morning and one in mid-afternoon. During these feeding peaks, they usually feed on fruit and seeds, leaves and petioles, sprouts, flowers and young branch tips. These food items can be found in a variety of forest types, including secondary forest, rubber plantations (both mature rubber and young plantations), and cultivated land. Their normal daily activity consists of resting, feeding, travelling and playing. Resting usually occurs during mid-day.

Conservation Status

Due to the lack of clarity on this species' taxonomy and distribution, *P. f. percura* is listed as Data Deficient in the IUCN Redlist.

Where to See It

The East Sumatran banded langur inhabits a variety of forest types, including secondary forest, and cultivated areas in Central East Sumatra, in a small area between the Rokan and Siak rivers. From Pekanbaru, Riau Province, you need to drive approximately 3-4 hours in order to find the good forest between the Rokan and Siak rivers.

32. Black-crested Sumatran Langur

Presbytis melalophos

Other common names: Simpai (Jambi, West Sumatra), Surili Sumatera (Indonesia)

Identification

Previously, several authors recognized four subspecies, *P. m. melalophos, P. m. sumatrana, P. m. mitrata,* and *P. m. bicolor,* but in

2014 Roos et al., recognized these subspecies as distinct species. This is still subject to debate among taxonomists. In general, the body and tail coloration are pale red, red orange or red brown and usually overlain with black. The outer side of the limbs is paler and more orange in color compared to the body, while the underside part of the tail is whitish. There is usually a single indistinct whorl on the forehead with a black crown crest. They are white, yellow or pale orange underneath. They also have flattened nape hairs that flank the central erect crest. This species has black facial skin and blackish to orange blackish eyebrows. Their cheeks are straw-colored, separated from a whitish forehead by a reddish or blackish band.

Geographic Range

They are found through out west Sumatra, from the upper Rokan River, just north of Gunung Talakmau, south to the upper Batang Hari River. They also range along the Barisan Range, west of Lampung Province.

Behavior and Ecology

More studies on this species are needed as information on their behavior and ecology is still very limited. As with other langurs, the main diet is fruit, seeds, flowers, young leaves and some insects. This species is known to consume more than 55 different species of plants. They occur in various tropical forest types. To move from tree branches they jump but they are also often seen to move quadrupedally if the branch is big enough. This langur is also capable of movement by brachiation to reach higher branches. Their home ranges can cover from 10 to 30 ha with average daily movements up to 1,300 m per day.

They are diurnal, with most activities occurring during the day. They are well adapted to different habitats. They can walk on the ground when no trees are available. They live in groups of uncertain composition. Sometimes one male is seen with several females and sometimes several males with several females. The group size can be from 8 to 12 individuals.

Conservation Status

They are listed as Near Threatened on the IUCN Redlist, as much of the range of this sub-species remains forested and it is unlikely to be declining fast enough to be considered threatened at this time. Conversion of habitat will be a problem in the future so if forest loss continues at current rates it could easily decline enough to qualify as Vulnerable.

Where to See It

This species prefers to live in primary lowland rainforests, hill forests, inland secondary forests, and sub-montane forests (Crockett and Wilson, 1980; Wilson and Wilson, 1976; Aimi and Bakar, 1996). They can be found in Kerinci Seblat National Park, Bukit Dua Belas NP, and Bukit Tiga Puluh NP. Sometimes they are also seen along the Trans-Sumatra way from Jambi to Palembang.

There are two good sites to see this species in Kerinci Seblat National Park, close to Bengkulu. From Bengkulu city in Southern Sumatra, you can drive toward Muara Aman for 4 hours or from Bengkulu to Argamakmur for 2 hours Kerinci Seblat National Park can also be reached by car from Padang, West Sumatra to Sungai Penuh, which will take 7 hours.

33. Black Sumatran Langur

Presbytis sumatrana

Other common names: Surili hitam (Indonesian), Simpai hitam (Sumatra)

Identification

The dorsal side and the outer surface of the limbs are dark brownish or brown in color and the ventral side and the inner surface of the limbs to the wrist and ankleare creamy white (Aimi and Bakar, 1992). The dorsal surface of the tail is brownish black and the ventral is creamy white (Aimi and Bakar, 1992). The hands and feet of this subspecies are black (Aimi and Bakar, 1992). The hair on the sides of the median crest is the same color as the back, which is colored dark brownish gray. The throat is creamy white, the face bluish, the muzzle flesh-colored, and the lips black (Aimi and Bakar, 1992).

Geographic Range

This species is found in Sumatra from north of Gunung Talakmau to the east side of Rokan river. They also occur on Kepulauan Batu (Pini island) in western Sumatra

Behavior and Ecology

More studies on this species are needed as information on their behavior and ecology is still very limited. Most activities occur during the day. The main diet is fruit, seeds, flowers, young leaves and insects. They can occur in a range of different tropical forest types. This langur moves by jumping from one tree to another and is also seen to move quadrupedally whenever possible. They are also capable of brachiation to reach higher branches. Home ranges can reach from 10 to 30 ha with average daily movements of up to 1,300 m per day.

They are well adapted to different habitat types including logged areas and plantations. In some habitats they are seen to walk on the ground when no trees are available. Black sumatran langurs live in groups of uncertain composition. Sometimes one male is seen with several females and sometimes several males with several females. The group size can be up to 8 to 12 individuals (Aimi and Bakar, 1992).

Conservation Status

This species is listed as Endangered in the IUCN Redlist (as *P. melalophos sumatranus*).

Where to See It

Black sumatran langurs can be seen in forests near the Batu Islands, North Sumatra. They also can be seen in Sungai Barumun, which is on the border between Riau Province and North Sumatra Province. The Barumun area is a tourist attraction located in Desa Batu Nanggar, Kecamatan Batang Onang, Kabupaten Padang Lawas Utara (Paluta). There are three ways to reach the Barumun area, which are:

- From Medan to Aek Godang then to Batu Nanggar (± 475 km, 8 hours drive)
- From Medan to Pematangsiantar then to Tarutung and to Aek Godang and finally to Batu Nanggar (± 510 km, 9 hours drive)
- From Medan to Kota Pinang via Gunung Tua then to Aek Godang and finally to Siondol (± 525 km, 11 hours drive)

34. Black-and-white Langur

Presbytis bicolor

Other common names: Surili hitam
putih (Indonesian), Simpai dwiwarnai

Identification

This species was previously known as sub-species *Presbytis melalophos bicolor* but in 2014 Roos et al., recognized it as a separate species, *Presbytis bicolor*. The color is dark gray-black or brown on the back and outer arms contrasting with a white underside, throat, limbs, inner surface of the tail, and the tip of the tail. The face is black with grey or bluish-grey circumocular skin and a black muzzle. The chin is grey or flesh-colored and there is a black fringe along the forehead (Groves, 2001).

Geographic Range

This species occurs in Central Sumatra, from near Sawahlunto and then east between Muaratebo and Rengat in the highlands from lower Sungai Batang Hari in the south to Sungai Indragiri in the north (Aimi & Bakar, 1992; Groves, 2001).

Behavior and Ecology

Information about this species is very limited due to its recent recognition as a distinct species, separate from *Presbytis melalophos*. In general, *Presbytis bicolor* is known to be folivorous but has also been seen to consume fruit, flowers, seeds, and some species of small insect. They are diurnal and live in groups of between 8 to 12 individuals consisting of both females and males. They are found in primary and secondary hill rainforest, shrub or secondary growth, and plantations. They also can be found in disturbed habitat and are considered to be quite tolerant of habitat changes.

This species moves quadrupedally in trees and uses a form of brachiation to swing from tree to tree. They are also seen on the ground where they access food and water. Based on *P. melalophos*, the home range is likely to be between about 14 and 30 ha.

Conservation Status

This species is listed under as Data Deficient in the IUCN Red list as *P. melalophos bicolor.*

Where to See It

You can see this langur in Bukit Dua Belas and Bukit Tiga Puluh National Parks in Riau in the area between Pekanbaru and Jambi Province. You can visit Camp Granit at Bukit Tiga Puluh National Park. From Pekanbaru, drive approximately 4 hours to Pematang Teba then 2 more hours to Camp Granit. From Jambi, the capital city of Jambi Province, you can drive 6 hours to Pematang Reba and 2 hours to Camp Granit.

35. Mitred Langur

Presbytis mitrata

Other common names: Lutung simpai putih (Indonesian), Chi-cha (Lampung), Kera putih (East Lampung)

Identification

Prebytis mitrata have a gray face with a pink muzzle and white or pink crescent shapes at the outer corner of each eye. The face is set off by a white cheek ruff and there is no whorl on the forehead. The overall body coloration is mouse brown or gray to very pale red-yellow gray, with some black overlay. The underside is creamy yellow or white, which extends up onto the flanks. The tail is redder than the body but the underside is a pale sandy color and it has a white tip. The limbs are between whitish and gray or red-brown with very pale-rooted hairs. Their hands and feet are gray (Groves, 2001).

Geographic Range

They are found in southern Sumatra, in Lampung, Palembang, and as far north as the Batang Hari River and west to the Barisan Mountains but not including those mountains (Groves, 2001).

Behavior and Ecology

More studies on this species are needed as information on their behavior and ecology is still very limited. Like other langurs, they are active during the day. The diet consists of fruit, seeds, flowers, young leaves and insects. They can occur in a range of different tropical forest types. They jump from one tree to another and also move quadrupedally whenever possible. They are also capable of using brachiation to reach higher branches. Home ranges are between 10 and 30 ha with an average daily movement up to 1,300 m per day.

They are well adapted to different habitat types such as logged areas or plantations. In some habitats they are seen to walk on the ground when no trees are available. Mitred langurs live in groups of uncertain composition. Sometimes one male is seen with several females and at other times there are several males with several females. The group size is between 8 to 12 individuals (Mittermeier et al., 2013).

Conservation Status

Mitred langurs are listed as Endangered in the IUCN Redlist, as *P. melalophos mitrata*, because at least 80% of their forest habitat was removed during colonial times. They are also very popular with the illegal pet trade.

Where to See It

This species can be found in Way Kambas National Park and Bukit Barisan Selatan NP, and in Kerinci Seblat NP to the south. The easiest way to see this species is to go to Way Kambas NP. As well as langurs, Way Kambas NP has many other species including Sumatran Elephants and Sumatran Rhinos. There is a Sumatra Rhino Sanctuary (SRS) that was established and is run by Yayasan Badak Indonesia (YABI). To reach this national park there are several options, such as:

Fly from Jakarta (Soekarno-Hata Airport or Halim Perdana Kusuma Airport) to Branti Lampung Airport, which takes about 30 minutes and flights are available daily. There are also frequent flights from Padang, Batam, and Medan. From Branti Lampung Airport to Labuan Ratulama (Way Jepara) using a rental car takes about 1.5 hours. An alternative is to use public transport, which takes between 2 and 3.5 hours via the east sumatra highway from Teluk Betung in Lampung City to Sribawono and to Way Jepara.

Visitors with private transportation such as a rental car can go directly to the Way Kambas NP gate in Plang Ijo while visitors using public transportation can rent a motor bike from Labuan Ratulama (Way Jepara) to Plang Ijo, which takes about 15 minutes.

36. White-fronted Langur

Presbytis frontata

Other common names: Puan (Sea Daya (Banjar, Sarawak)

Identification

The body is pale grayish brown with a yellowish gray tail. The chin and lower cheek are greyish but the head, crest, brow, hands, feet and cheeks are blackish (Rowe, 1996; Groves, 2001). There is a distinct white-colored roughly triangular crest on the forehead. The White-fronted langur has a tall sagittal crest that leans forward. Coloration of the newborns is not known.

The average body weight of males is 5.67 kg and females, 5.56 kg (Fleagle, 1999). They have a sacculated stomach to help breakdown cellulose from their highly folivorous diet. They also have an enlarged salivary gland. Their face is short and broad and their jaw is deep. This species has narrow incisors and sharp molars.

Geographic Range

This species occurs in both Malaysian and Indonesian Borneo. It is found in central and eastern Borneo, with fewer populations in the west. In Eastern Kalimantan, they range from Sungai Kayan and Sungai Segah southwards, including the Kutai reserve, Sepaku, inland of Balikpapan and to the coast of South Kalimantan. Then west to Long Petah, Sungai Kayan, the upper Sungai Mahakam, Puruk Cahu in the upper Barito River and in the vicinity of Banjarmasin.

Behavior and Ecology

There is limited information on the habitat and ecology of the White-fronted langur. Nevertheless, this species is known to be predominantly folivorous, but has also been observed to consume fruit and seeds. Mostly, they prefer young leaves over mature ones. During their active time in the day, they move by brachiation and like other langurs, they are able to leap between trees easily. They are also found occasionally on the ground.

They are polygynous and are found in small groups. Group size ranges from 2 to 10 individuals (Suzuki, 1984). Young males may be solitary after leaving their natal group. Population estimates vary in different areas. It is considered to be common in the northwestern Malaysian mountains (Lanjak-Entimau, Bantang Ai and Bentuang Kerima), but uncommon in the other parts of its range. This may be a sampling bias due to the fact that they are shy where hunted and generally difficult to see (IUCN, 2015).

Conservation Status

The White-fronted langur is listed as Vulnerable in the IUCN Redlist (IUCN, 2015) and included in CITES Appendix II. They are threatened by hunting for meat and traditional medicine, as well as by habitat loss. Detailed ecological and conservation studies of this leaf monkey are not currently available from the Kalimantan's rainforest. This lack of available information may be a reason that no protection plan has been developed for this endemic leaf monkey. Loss of habitat through logging is assumed to be the main threat to their existence in the wild. Although this species is legally protected in Indonesia, this level of protection is insufficient. Essentially, their wild population is not yet seriously threatened. Nevertheless, this species will become threatened if no conservation plan is put in place.

Where to See It

Typically, they are found in primary lowland dipterocarp forest, such as in Kutai National Park in East Kalimantan or in the forests in the Pleihari area of South Kalimantan. They can also be found in the Meratus forest in South Kalimantan. This species appears to be limited to less than 300 m above sea level. They have also been observed in riverine and hill forest, as well as occasionally in secondary forest and plantations (Azuma et al., 1984). If you choose to go to Kutai National Park, you probably can see other endemic and rare primates such as *P. canicrus*. To go to Kutai National Park you need to fly to Balikpapan of East Kalimantan province. From Balikpapan, you can fly to Bontang by small plane or drive there for approximately 6 hours. From Bontang to Sangkima is a 1 hour drive, the closest village.

37. Maroon Langur

Presbytis rubicunda

Other common names: Lutung merah (Malay); Kalahi, Pampulan (Banjar), Jalur merah (Sea Dayak), Khallasie (Kayan), Kalasi (Ngaju), Kelasi (Gunung Palung), Kera merah (Karimata island)

Identification

This species includes five sub-species, four of which occur in Indonesia, the Maroon Red Langur (*P.r. rubicunda*), the Red-naped Red Langur

(*P.r. crimatae*), the Orange-naped Red Langur (*P.r. ignita*), and the Southwest Kalimantan Red Langur (*P.r. rubida*). One sub-species, the Orange-backed Red Langur is found in Sabah, Malaysia (Roos et al., 2014). The head and body length is between 44 and 58 cm for males and between 48 and 52 cm for females, with tail length around 65 cm. The maximum weight can be up to 7 kg. *Presbytis rubicunda* have unique distinct hair on the forehead which is arranged in a radiating manner. Generally, they are born white but soon change color to a striking maroon-red to brick red with a lighter underside and blue-gray facial skin, except for the upper-lip and chin, which are pinkish. Groves (2001) mentions that sub-species in the east of their range are distinguished from those in the west by body color, and sub-species in the south from those in the north, on the basis of a narrow-crest versus a broad-crest.

- *Presbytis rubicunda rubicunda* (Maroon Red Langur)
  This subspecies has maroon to mahogany red body color with paler underside and inner surface of the limbs. The hands and feet, especially on the digits, are a heavily washed black color. They have 'narrowed-crested' hair on the forehead.
- *Presbytis rubicunda rubida* (Southwest Kalimantan Red Langur)
  Compared to Presbytis rubicunda rubicunda, this is more yellow in color, while the hands and feet are not or only somewhat blackened. They have 'narrowed-crested' hair.
- *Presbytis rubicunda ignita* (Orange-naped Red Langur)
  This sub-species is fox red in color with no blackened hands and feet. It displays a rather 'broad-crested' hair crown.
- *Presbytis rubicunda carimatae* (Red-naped Red Langur)
  Compared to the other sub-species in this one is a dark brick red with slightly darkened hands and feet. It has a 'narrow-crested' hair crown with golden tinge.

Geographic Range

Presbytis rubicunda is endemic to Borneo island and the adjacent Karimata Island (Nijman & Meijaard, 2008). *P. rubicunda rubicunda* is distributed throughout mainland Kalimantan ranging from sea level to 1,200 m. *P. rubicunda rubida* occurs in the Southwestern part of Borneo, south of Sungai Kapuas and west of Sungai Barito. *P. rubicunda ignita* can be found in Northwestern Kalimantan from Sungai Kapuas, in both the lowlands and highlands and extending to Sarawak and the borders of Brunei. *P. rubicunda carimatae* is found only on Karimata Island, West Kalimantan (Groves, 2001).

Behavior and Ecology

Presbytis rubicunda is a folivorous species, but could also be considered as granivorous due to the high amount of seed consumed in certain months of the year. The diet consist mainly of seeds and young leaves, but sometimes they also consume fruit and buds as well as insects (Ehler Smith et al, 2013b). They acquire fruit located at the tip of tree branches by bending the branch and then reaching it by mouth if the fruit is small or hands if large (Supriatna et al., 1986). Some fruits are eaten entirely (e.g. genus *Fragraea*) while of other species, only the flower petals are eaten (genus *Deplanchea*). The subspecies *Presbytis rubicunda carimatae* was found to consume more leaves than other sub-species (Yanuar et al., 1993). Interestingly, this species also occasionally practices geophagy, the eating of soil, some from termite-mounds (Davies & Baillie, 1998). According to Davies and Baillies, this soil-eating behavior benefits the animal by easing digestive disorders such as fore stomach acidosis as well as providing supplementary minerals.

This species is active during the day and mostly arboreal, employing both brachiation and jumping in moving through trees. Occasionally, they are seen on the ground. They prefer the tree canopy, which is assumed to be to avoid humans or predators (Bersacola et al., 2014). The Maroon leaf monkey appears to be territorial with home ranges between 105 and 112 hectares (Ehlers Smith et al., 2013a). Groups overlap their home ranges, but each group has a core area where the sleeping site is located (Suprianta et al., 1986). The size of a group's home range is correlated with the size of the group; smaller groups have smaller home ranges (Supriatna et al., 1986). Group sizes of the Maroon leaf monkey range from 2 to 7 individuals (Davies, 1984; Supriatna et al., 1986; Davies and Baillie, 1988).

They are polygamous with only one male in each group (Suprianta et al., 1986; Davies, 1987). They only have one adult male per group (van Schaik and Horstermann, 1994). Shortly before adolescence the males emigrate from their natal group, and then usually travel alone or in an all-male group before they build their own unimale group (Suprianta et al., 1986).

Conservation Status

This species is threatened by the loss of its forest habitat. On Karimata island, human population increase and the accompanying increase in gardens and field cultivation threatens the subspecies *P. r. carimatae*. Maroon leaf monkeys are also threatened by hunting for meat and traditional "medicine" (IUCN, 2015). Nonetheless, this species is moderately adaptable to antropogenic disturbance and can occur in secondary habitats. They are listed as of Least Concern in the IUCN Redlist.

More detailed studies on the ecology and conservation of all sub-species are needed in order to develop strategies for their conservation in the wild.

Where to See It

Maroon leaf-monkey are relatively well adaptable to human disturbances. They can tolerate forests that have been regrown after logging. They are usually found in both primary and secondary lowland forests below 2,000 m above sea level. They are also found in swamp forests (Chivers and Burton, 1988). On Karimata Island the subspecies *Presbytis rubicunda carimatae* prefers to live in swamp forests and occasionally visit native gardens in search of food (Yanuar et al., 1993). The best site to see this species is in Tanjung Puting National Park, which I visited many times. You can fly to Pangkalan Bun with a small plane from Jakarta or Banjarmasin or Pontianak or Ketapang. From Pangkalan Bun, you may drive to Kumai (15 km) then go by speed boat (30 minutes) or by big boat (Klotok in bahasa) for 3 hours to Camp Leakey.

38. Hose's Langur

Presbytis hosei

Other common names: Lutung banggat (local)

Identification

Hose's langur is grey on the upperpart of the body and white underneath. The hands and feet are a blackish color. Facial skin (lower jaws and cheeks) is pinkish with a distinct black band across each cheek (Payne et al., 1985; Groves, 2001). The tip of the crown is black, and sometimes the rest of the crown is also black (Groves, 2001). Infants have a white body with black lines down the back and across the shoulder (Payne et al., 1985).

Geographic Range

Presbytis hosei is found only on Borneo, in the countries of Indonesia, Malaysia and Brunei Darussalam (Nijman, 2010). In Indonesia, it is restricted to the northern parts of Kalimantan in North Kalimantan Province. The species is widely distributed from Kayan Mentarang, to the Bahau river, the Karangan river, and Kutai Barat regency (Groves, 2001; Nijman, 2005). Its exact distribution in Kalimantan is still not accurately known.

Behavior and Ecology

Ecologically, *Presbytis hosei* is very similar to the maroon leaf monkey. They are arboreal primates moving quadrupedally and by leaping (Fleagle, 1988). They are frequently found in tree canopies feeding, resting, jumping, travelling and ranging during the day. Once in a while they are also seen climbing down to the ground to visit natural mineral sources (Payne et al., 1985).

    P. hosei is primarily folivorous. The diet mostly consists of leaves and seeds, but they also consume fruit, buds and flowers. Eggs and nestlings of the gray-throated babbler, *Stachyris nigriceps* have also been seen to be consumed by this species (Goodman, 1989).

    They are polygamous with one adult male and two or more adult females in small family groups of 6 to 8 individuals (Payne et al., 1985). Solitary individuals do occur (Bennett and Davies, 1994), usually young males that have left their natal group (Rowe 1996; Bennett and Davies, 1994). This species is known to form mixed-species associations with the maroon leaf-monkey, *Presbytis rubicunda* (Rowe, 1996).

Conservation Status

P. hosei is listed as Vulnerable in the IUCN Redlist. A seven year long census of this species in Kayan Mentarang National Park, Kalimantan, found that density had dropped by 50-80% due to hunting for traditional medicine (Nijman, 2005). They are hunted for their gallstones as well as for food (Rowe, 1996). Apart from hunting, forest decline has led to the loss of habitat and caused a loss of food sources, canopy trees for ranging, and night-sleeping trees.

Where to See It

P. hosei can be found in primary and secondary forest. They also occasionally enter local plantations and logged forest (Payne et al., 1985). This species is found between 1,000 and 1,300 m, in altitude in the hill ranges (Goodman, 1989). Patience is needed when looking for this species because they are often hiding and can only be heard. They are also rare and therefore to find them you need to hike a little away from the main road. To go to Kayan Mentarang National Park, you need to fly from Jakarta to Tarakan (3 hrs). From Tarakan you can fly with a small plane to Malinau (30 mnts) then to Long Alango (1 hour). In Malinau Regency, there is a famous Kenyah dayak tribe who have a traditional forest called Tanah Olen (more than 5000 ha) where you can see this Hose's langur. You need to drive from Malinau to Setulang approximately 1-2 hours or by boat for 2 hours.

39. Sumatran Grizzled Langur

Presbytis thomasi

Other common names: Lutung Rungka, Kedih (Southeast Aceh), Kek-kia, Thomas's leaf monkey

Identification

Presbytis thomasi is gray on the back and limbs and white underneath (Wilson and Wilson, 1976). The tail is pale colored on the ventral side and gray on the dorsal side (Wilson and Wilson, 1976; Aimi and Bakar, 1992). The hands and feet are black (Aimi and Bakar, 1992). There is a blackish vertical stripe on the crest of the head with white patches on either side of the crest (Wilson and Wilson, 1976). The face is light gray with a blackish colored mustache (Wilson and Wilson, 1976). The chin is white and the muzzle is flesh-colored. (Aimi and Bakar, 1992). The infants of this species are white (Wilson and Wilson, 1976). Adult males have an average body mass of 6.67 kg and females 6.69 kg (Fleagle, 1999).

Geographic Range

Thomas's leaf-monkey is found on the island of Sumatra, Indonesia. It occurs in North Sumatra, north of Lake Toba, and southeastern Aceh. There are substantial populations in the Bohorok region and Ketambe-Gunung Leuser National Park. In Aceh Province, they are found north of the rivers Simpangkiri and Wampu (Wilson and Wilson, 1976). Their range has recently been extended to the south of Simpangkiri River (Aimi and Bakar, 1996). This species is sometimes found in rubber plantations (Gurmaya, 1986). It can usually be found in primary and secondary forests (Gurmaya, 1986; Wilson and Wilson, 1976).

Behavior and Ecology

Thomas's leaf-monkey is primarily folivorous (Ungar, 1995), although in some parts of its range it may be more frugivorous (Gurmaya, 1986).

The diet consists of fruit, young leaves, seeds flowers, roots and tubers, especially the soft, sappy and sticky portions of plants. They are also known to consume insects and soil. They are especially fond of the softer leaves of rubber trees. The peak feeding times are in the morning and late afternoon (Ungar, 1995).

Fruits consumed by this species tend to have high pH levels, be unripe, are large, and have hard rinds or husks (Ungar, 1995). Examples include: *Gnetum cf. latifollum, Paranephelium nitidum,* and *Quercus* sp. (Ungar, 1995). Large seeds of dry fruits of the species *Dysoxylum spp., Cnestis platantha,* and *Scorodocarpus borneensis* are consumed by Thomas's leaf-monkeys (Ungar, 1995). Fruits with high tannin levels are consumed because the fore-stomach microbes are able to break down the tannins (Ungar, 1995). They will avoid fruits with fleshy sugar-rich pericarps having a low pH because the bacteriocidal resins in the pericarps would kill fore-stomach microbes and a low pH level could cause acidosis of the fore-stomach (Ungar, 1995). Gurmaya (1986) found that the following species were the staple food for Thomas's leaf-monkeys: *Samanea saman, Hevea brasiliensis, Mikania chordata, Artocarpus integer, Sandoricum koetjape, Musa sp.,* and *Durio zibethinus.* They use the front teeth to pierce through the hard husks of fruit or through thick bunches of leaves (Ungar, 1996).

Occasionally, they eat toadstools and the stalks of coconuts (Gurmaya, 1986). This species has also been seen feeding on gastropods such as ground snails (Steenbeek, 1999) and insects and other small animals. Most water is obtained from the food consumed, but individuals have been observed to drink water from tree holes and small streams (Gurmaya, 1986).

Generally, Thomas' leaf monkey will choose a night-sleeping site in the open situated near human paths, plantations, roads, river banks or farmlands. Typical night-sleeping trees are high trees with long branches, thin foliage and free from creepers. In behavior, habitat preference and relative population size it is very similar to the Mitred leaf monkey.

Group size varies from 3 to 21 individuals (Gurmaya, 1986). This is a semi-arboreal, diurnal species (Gurmaya, 1986). Individuals do occasionally come down to forage for mature rubber seeds and durian fruit (Gurmaya, 1986). Ungar (1996) found that this species will feed on the ground on average once per day.

Conservation Status

Habitat destruction through logging operations and conversion of forests for agriculture, estate crops and human settlement are major threats to their existence in the wild. This species has been legally protected by Indonesian law. They are listed under CITES appendix II and as Vulnerable on the IUCN Redlist.

Where to See It

They can be found in primary and secondary lowland rainforest, swamp forest, lowland alluvial forest, fruit plantations and rubber tree plantations. Thomas' langurs are relatively easy to spot in their natural habitat. You can go to Gunung Leuser National Park and other forests in Aceh (e.g. Janto Protected Area). Sometimes they can also be seen on forest edges or local plantations. The easiest place to access to see this primate is in Sibolangit Education Center, which is around 30 km from Medan, North Sumatra in the direction of Kabupaten Kaban Jahe. This area is less than 1,000 ha. Walking trails provide access and hotels are available in the area surrounding Sibolangit.

40. Pagai Langur or Mentawai Leaf Monkey

Presbytis potenziani

Other common names: Lutung
Mentawai (Indonesia), Ataipeipei
(Sipora, North and South Pagai)

Identification

Pagai Langurs show minimal sexual dimorphism (Tilson and Tenaza, 1976). The average body weight of adult males is 6.5 kg and adult females, 6.4 kg (Fleagle, 1999). Head and body length is around 50 cm with a tail of around 58 cm. Pagai Langurs are jet-black above and on the tail with reddish-orange undersides and inner sides of thighs. Inner sides of the upper arms are whitish or reddish and the throat, cheeks, and chin are yellowish-white grading backward through gray to the reddish color of the underside. They have a whitish brow and black facial fur except the mouth because it is somewhat depigmented around the mouth,. Newborn infants are totally white. After a few days the pelage of the infant's face becomes darker, and after two to three weeks the entire pelage darkens beginning at the dorsal midline and the head (Tilson, 1976). The belly and chest turn a reddish-brown, the back becomes black, and several months later only a fringe around the face and the tip of the tail remain white (Groves, 2001).

Geographic Range

This species can be found only on the Mentawai Islands of Sipora, North Pagai, South Pagai and nearby Sinakak.

Behavior and Ecology

Presbytis potenziani are normally active during the day and move by semi brachiation. They are good jumpers among trees in the canopy. In one study, they were observed to move in the early morning to primary forest to feed during the day time, and late in the evening return to the sleeping tree which is usually located outside the primary forest, usually in coconut groves or swamp forest. This movement pattern is probably a way to avoid competition in feeding and for sleeping trees with the sympatric Kloss's Gibbon.

This species is highly frugivorous for a colobine but the relative proportions of fruit and seeds in their diet seems to vary depending on habitat. In primary forest habitat on North Pagai, fruit and seeds constituted 32% of their diet while in the secondary forest habitat, fruit and seeds contributed about 70% of their diet. They also eat flowers, sap, bark, and fungi and they are one of only a few primates that can eat the leaves of the dipterocarp tree family.

The density varies in each island from 0.5-5.1 individuals/km2 (Paciulli, 2010).Yanuar and Supriatna (2018) estimated that their density varied from 2.8 groups/km2 in Sipora and 2.7 groups/km2 with standard deviation of 0.9. The average group size of this langur is 3.8 in Sipora and 3.8 in Pagai.

Pagai Langurs can be found in unimale-unifemale, unimale-multifemale, and multimale-multifemale groups (Watanabe, 1981; Tenaza, 1987; Fuentes, 1996). Both males and females disperse from their natal group, with the females tending to leave at adulthood (Fuentes, 1996, 1977a). The ranges of groups will overlap, but a group will stay out of a neighboring group's core area (Watanabe, 1981; Fuentes, 1996). If a neighboring group comes into another's core area then the resident adult male and the intruding male will engage in aggressive, loud vocalizations and visual displays (Tilson and Tenaza, 1976). Members of the group will remain close together during feeding, resting, and travelling (Fuentes, 1996). Pagai Langurs will respond to alarm calls given by Kloss' gibbon, *Hylobates klossii* (Tenaza, 1987). They are also found to be subordinate to Kloss' gibbons, being supplanted from feeding areas when approached (Tilson and Tenaza, 1982). Infant parking has been reported in this species where a female will put down an infant and then forage (Fuentes and Tenaza, 1995). Pagai Langurs have been found to form associations with pig-tailed langurs, *Simias concolor*, and Mentawai pig-tailed macaques, *Macaca pagensis* (Rowe1996).

Conservation Status

The Mentawai primates are entirely dependent on forests that are now critically threatened by legal and illegal logging, forest clearance and resource extraction. They are hunted by local people for food and for the illegal pet trade. Much of the forest is highly disturbed due to industrial-scale logging by companies present on all four islands. Consequently, habitat for the primates on the Mentawai Islands has decreased by at least 50% in the last 25 years (Chivers, 1986).

P. potenziani is listed as Critically Endangered in the IUCN Redlist and listed in Appendix I of CITES. They are also protected under Indonesian law. Similar to the Siberut Langur, threats to this species are logging and hunting. Local people are reported to climb their sleeping trees at night to shoot them. Hunting has increased recently because roads now provide access to remote areas and bows and arrows have been replaced by riffles. Local taboos and rituals that regulated hunting in the past have been lost with the advent of Christianity. They also are sometimes kept as pets (Fuentes, 1997b).

Where to See It

They can be seen in primary and secondary lowland rainforest, swamp forest, logged forest, presumably mangrove forest, and cultivated areas. They spend most of their time in the middle and upper canopy. They are also sometimes seen coming to the ground. There are only several places to easily observe this primate on North and South Pagai islands and on Sipora Island, in the Mentawai Islands. To go to North and/or South Pagai, or Sipora, you need to go by ferry from Padang, West Sumatra Province. It will take between 6 to10 hours depending on the weather and the type of boat. On Sipora island, you can go to Saurienu, situated in north-central Sipora, about 15 km from Tua Pejat, the capital city of Mentawai regency. The neighboring primary forests around Saurienu, such as the Siberimanua, Tua Pejat, and Betumonga areas, have been logged intensively since the 1970s. In South Pagai, only 900 km2 of forest remains, and much of it has been affected by logging since the 1970s. The forests are predominantly old secondary growth with relatively little pristine old growth lowland forest (Yanuar and Supriatna, 2018).

41. Siberut Langur

Presbytis siberu

Other common names: Lutung Joja siberut (Indonesian), Joja (Siberut)

Identification

Originally this species was named *Presbytis potenziani* but now has considered to be a separate species. As in the Pagai langur, there is minimal sexual dimorphism in this species. Males weigh around 6.5 kg and females around 6.4 kg. In general, head and body length is between 48 and 50 cm with a tail between 58 and 64 cm. Adults are predominantly black with a whitish tuft around the face, the top of their head, around the neck, and in the pubic region in both sexes. The ischial in males is connected but separate in females. Males have a white scrotum. Facial skin is black but becoming somewhat white around the mouth. Coloration is similar to the Pagai Langur, but Siberut langurs are darker and the white pubic patch is more sharply demarcated (Mittermeir et al., 2013).

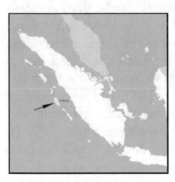

Geographic Range

They are only found on Siberut Island in Mentawai Islands

Behavior and Ecology

Ecological and behavioral information on this species has been known from long term research carried out by the

German Primate Center on Siberut island. This species is arboreal and active during the day. The size of its home range is between 11.5 and 22 ha, which is smaller than that of the Pagai langur. Only unimale-unifemale groups have been observed, unlike the variable social organization found in the Pagai langur. A population density of 8.2 individuals/km^2 is reported in northern Siberut. They are sympatric with the siberut macaque (*Macaca siberu*), the pig-tailed langur (*Simias concolor*), and kloss' gibbon (*Hylobates klossi*) (Mittermeier et al., 2013).

Conservation Status

Siberut langurs are listed under CITES Appendix I and as Endangered in the IUCN Redlist. The Siberut Langur is threatened by logging and is a popular target among Mentawai hunters. Similar to other primates in Mentawai, hunting has increased recently because roads now provide access to remote areas and there has been a shift from bows and arrows to rifles. Local taboos and rituals that regulated hunting in the past have been lost with the advent of Christianity. They are also sometimes kept as pets.

Where to See It

They can be found in primary and secondary lowland rainforest, swamp forest, logged forest, possibly mangrove forest, and cultivated areas.

To see this species, it is recommend that you contact researchers who are currently studying it. You can go to the national park office in Padang and ask.

Access to Siberut:

- Padang ⟶ Muara Siberut/ Muara Sikabaluan in Siberut Island (± 155 km) can be reached in about 10 hours using the ferry boat, which operatesthree times a week. From Muara Siberut/ Muara Sikabaluan ⟶ Siberut NP can be reach using a motor boat on the rivers.

In Siberut, there is accommodation available at the Syahruddin Hotel and the Wisma Tamu guesthouse. Local guides are avalable in Padang, Bukittinggi and Muara Siberut. To find local guides you can check in Jl. Pontianak N/13 Padang, West Sumatra.

42. Natuna Island Langur

Presbytis natunae

Other common names: Lutung natuna (Indonesian), Kekah (Bunguran)

Identification

Presbytis natunae has similar coloration to *Presbytis siamensis*, except for the very distinct white colored cheek whiskers (on the upper margins) that are exceptionally bushy and prominent. They do not possess a forehead whorl or the whorl is indistinct (Groves, 2001). That is why previously Brandon Jones (1984) considered this monkey to be a subspecies of *Presbytis siamensis*. The body is dark grayish brown. They have a darker head and lower parts of limbs with a white-colored underside, which extends to the back of the thighs, chest, chin, and often to wrists and ankles but not under the tail. Hairs on the chest are directed outward, except on the midline part. They have very large white-colored eye rings (especially below the eye), but the upper and lower halves of the eye rings do not actually connect around the eyes. (Groves, 2001; Mittermeier et al., 2013).

Geographic Range

Presbytis natunae is found on Bunguran Island and Natuna Island, in the northern part of the Natuna Islands, off the northwestern coast of Borneo, Indonesia (Nijman & Lammerlink, 2008).

Behavior and Ecology

This species is known to be active during the day and arboreal. Behavior and ecology of the Natuna Island langur is still unknown.

Conservation Status

The Natuna Island regency has not yet established any parks or protected areas to help conserve and manage its natural resources. As a result, the Natuna Island langur has been ranked among the 25 most endangered primates in the world (Primate Specialist Group SSC-IUCN, 2000). In late 2000, approximately 41,000 ha of natural forest were identified for conversion into oil palm by the companies PT. Rupat Sawit Lestari and PT. Sinar Alam Sejati. Such a large-scale development would pose serious threats to the last remaining habitat of this island's endemic species.

Where to See It

This species is known in various habitats. They can be found in primary forest, logged forest and rubber plantations. To reach the Natuna Islands you can take a flight to Batam Island and then take a small plane to Bunguran Island. From surveys in 2008, in some places there is still some forest left in this island where you may see this species.

43. Cross-marked Langur

Presbytis chrysomelas

Other common names: Lutung ceeka kalimantan (Indonesian), Nokah (Pontianak)

Identification

There are two sub-species of *Presbytis chrysomelas*, *P. c. chrysomelas* and *P. c. cruciger* (Roos et al., 2013). There is an undescribed sub-species in an isolated population found only in southeastern Sarawak (Groves, 2001). Groves (2001) described *P.c. chrysomelas* as having a jet black color with some brown hairs in the dorsum. They have a brown gray to white cheek with a pale zone on the underside that extends to cheeks, chin, wrist, halfway down thighs and underside of the tail. In his book, Groves also noted that there might be a variability in this sub-species.

Geographic Range

Presbytis chrysomelas is only found on Borneo island in Brunei, Kalimantan, Sabah and Sarawak. Only the subspecies *P. c. chrysomelas* occurs in the Indonesian part of Borneo, where it can be found in northwestern Kalimantan, northeast of the Kapuas River, to the borders of western Brunei and western Sarawak (Groves, 2001; Vun et al., 2011). It has been found in Betung Kerihun National Park and also Danau Sentarum National park.

Behavior and Ecology

The taxonomy of this species is complex and disputed. It has been considered by different authorities and at different times as a subspecies of *P. femoralis* or *P. melalophos*. There is no intensive study on the behavior of this species, but one study on movement patterns of *P. c. chrysomelas* in the Samunsam Wildlife Sanctuary of Sarawak, Malaysia showed that their ranging pattern and range may be influenced by the distribution and abundance of food resources. Group size is also assumed to influence daily movement (Ampeng & Md-Zain, 2012).

Conservation Status

In the early 20th century, this species was considered to be common (Baccari, 1904; Banks, 1931), in areas where today it no longer occurs. *P. c. chysomelas* occurs in less than 5% of its historic range. There are only five sites with recent records: Maludam National Park; Samunsam Wildlife Sanctuary; Similajau National Park; Tanjung Datu National Park, all in Malaysia and the Lingga area of Sarawak, also in Malaysia, although it has been found in Betung Kerihun National Park in Indonesia (Information from Betung Kerihun National Park based on pictured analyzed in 2018)

 P. c. chrysomelas is listed as Critically Endangered in the IUCN Redlist and is protected under CITES Appendix II. Combined population estimates from known locations are very low (approximately 200-500 individuals). Habitat conversion for plantations, especially for Palm oil is the main threat to this species, resulting in its disappearance from most of its former range.

Where to See It

You can find this species in swamp forest, lowland forest and mangrove habitat. The Cross-marked langur can be found in several areas in

West Kalimantan such as Gunung Nyiut Nature Preserve (180,000 ha) and Gunung Raya Pasi Nature Preserve (3,700 ha), even probably you can find them in Betung Kerihun National Park and Danau Sentarum NP. Gunung Nyiut Nature Preserve is rich in biodiversity. If you want to see Danau Sentarum National park, you can fly from Jakarta to Pontianak (1 hour) then you go to Kapuas Hulu Regency (1 hour 30 minutes). From Kapuas hulu you can drive to Danau Sentaraum National Park in about 3 to 4 hours.

44. Miller's Langur

Presbytis canicrus

Other common names: Lutung beruban kalimantan (Indonesian), Lotong beruban (Kalimantan)

Identification

This species was listed as a sub-species of *P. hosei* (*P. h. canicrus*) but Ross et al., recognized *P. canicrus* as a new species in 2014. The overall color is brown with a grayish underside anteriorly and a white belly and inner side of limbs. The facial skin is completely black down to a whitish line between the mouth and ear. P. canicrus exhibit sexual dimorphism with females bigger than males. The total body length of

females is 122.8±5.7 cm, and of males 118.1±3.2 cm).

Geographic Range

This species has a very restricted distribution along the east coast of Kalimantan from Kutai north to Gunung Talisayan (Groves, 2001). It was considered extinct, but recent surveys have confirmed populations in the Baai river area in East Kalimantan (Setiawan et al., 2009) and Wehea forest in East Kalimantan (Lhotha et al., 2012).

Behavior and Ecology

This species resembles *P. hosei* ecologically. They move quadrupedally or by leaping through the tree canopy. They feed, rest, and travel in the canopy, but occasionally climb down to the ground to reach salt-licks. The diet is mostly leaves and leaf shoots with a considerable amount of seeds. They also feed on flowers, buds, and insects.

They have a uni-male social system and a polygamous mating system (Payne et al., 1985). On average, groups consist of 8 individuals. Solitary males are usually young males that have dispersed from their natal group (Rowe, 1996; Bennet & Davies, 1994).

Conservation Status

Miller's Langur was considered to be extinct for a long time as it's last stronghold, Kutai National Park, was largely destroyed by illegal logging, illegal settlements, industrial development close to the park and fire, leaving an estimated 5% of the forest intact (Meijard and Nijman, 2000). But in 2012, researchers confirmed the existence of a population in Wehea forest, although its viability at this site is still unknown (Lhotha et al., 2012).

Reassessment of their status in the IUCN Redlist is needed. Currently, it is listed as Endangered under *P. hosei canicrus*. Ongoing habitat destruction and heavy hunting are the primary threats. Wehea Dayaks have invoked local custom in order to protect the habitat of this species. The Wehea Forest has been placed under "adat" law that prohibits tree cutting, fire lighting, and the harvesting of animals and plants.

Where to See It

This species is found within lowland dipterocarp forest to mountain forest (Lhota et al., 2012). They usually occupy primary forest and secondary forest, but may also be found in logged forest. They may also be seen visiting "sepan" or salt springs. Due to the rarity of this species, the most likely place to see them is in the forest near Sungai Baai in Wehea forest in East Kalimantan. This is a remote area so considerable effort would be needed to find it. To get there you need to travel to Balikpapan from Jakarta. From Balikpapan you can rent a car and drive to Muara Wahau. From Muara Wahau you can use a public ferry up to the Mahakam river and canoe upstream to Wehea approximately all day trip. But you can also take a plane from Balikpapan to Berau Regency then drive to Wehea forest.

45. Ebony leaf monkey

Trachypithecus auratus

Other common names: Lutung, Budeng (Jawa), Petu, Hirengan (Bali).

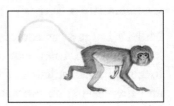

Identification

The pelage is black to silvery-black though a red color morph has been observed coexisting with the black. There is a slight brownish tinge on the underside, sideburns, and legs. Facial skin, palms and soles are black in the common morph, and depigmented or freckled in the red morph. Females can be distinguished from adult males by a pale, cream patch in the pelvic region. Infants have an orange coat.

Geographic Range

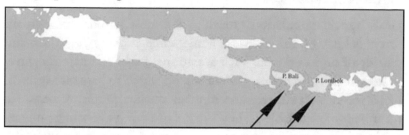

Common and widespread throughout Java, Bali, and reaching to Lombok.

Behavior and Ecology

Previously, Groves (2001) recognized two subspecies, *T. auratus auratus* and *T. auratus mauritius* but in 2014, Ross et al., updated the taxonomy to two different species, *T. auratus* and *T. mauritius*. *T. auratus* are active during the day. They move by semi-brachiating in trees canopies but sometimes descend to the ground.

Mostly, they consume young leaves, fruit, and flowers. Particular portions of food items are eaten e.g. seeds of some fruit, the calyx of some flowers and the petioles of some young leaves. Their food items tend to be low in fiber content and are more digestible than vegetation parts that are not eaten. Phenolics and condensed tannins are not a main determinant in food selection. Young teak leaves are the single most common food item of this leaf monkey.

The social system in this species is polygamous consisting of one or two or more adult males with several adult females and their offspring. Group size varies between 6 and 21 individuals with a mean of 10 individuals. Home range size is from 5 to 8.5 ha (Nijman and Supriatna, 2008).

Conservation Status

Java's human population growth is the most serious threat to the forest habitat of this species. While some primary forest remains outside of protected areas such as national parks and nature reserves in Java, little information is available on which to base management plans for this species either within or outside protected areas. No information is available on how successful a captive breeding program might be. Furthermore, this species is hunted for the pet trade and it is considered to be an agricultural pest. It is listed as Vulnerable on the IUCN Redlist.

Where to See It

This langur occupies a wide range of habitats from dry lowlands to high elevations, and in lowland swamp forest and mangroves and in secondary forest and plantations such as teak, mahogany and acacia. In general, they still can be seen in most national parks in Central and East Java, Bali and Lombok including Baluran NP, Meru Betiri NP, Alas Purwo NP, all on Java, Bali Barat NP, Bali, and Gunung Rinjani NP on Lombok.

46. Silvered Langur

Trachypithecus cristatus

Other common names: Cingkau (Sumatra), Buhia (Kalimantan), Bukis (Ngaju), Peut (Manyaan), Bochies (Dayak), Balung, Leso, Muis (Kapuas).

Identification

Groves (2001) recognized two subspecies of *Trachypithecus cristatus*, *T. c. cristatus* and *T. c. vigilans*, which are said to be genetically similar but *T. c. vigilans* is found only on Natuna Island. This species can be distinguished by a small head (they have relatively small teeth and jaws), large body size, stance, and a long tail hanging straight down as it sits on a branch of a tree. Generally, the Silvered Langur is lighter and grayer, with long gray hair tips compared to the Ebony leaf monkey. Circumfacial hair is also found in this species, but with a distinct pointed crest pattern that is noticeably different from *T. auratus*. The skin pigment is dark. The newborn pelage color is orange with white face, hands and feet. Infants darken with age and it takes 3 to 5 months to change completely to the adult coloration. First the head starts to fade to a yellow orange, then the head, face, hands, feet and ears turn dark gray and then black.

Head and body length is between 46.5 and 56.0 cm, with a 68.0 cm tail. The mean body weight for males is 7.1 kg while female weight is approximately 89% of the males' weight. Sexual dimorphism is slight, adult males are somewhat larger than adult females and the latter have irregular white patches on the inner part of the upper thigh. The sub-species each have their own pelage coloration:

* *T. cristatus cristatus*

  This sub-species is gray in color with long lightened hair tips evenly scattered over the body and crown. There is a rare red morph on Borneo.

* *T. cristatus vigilans*

  Compared to *T. cristatus cristatus*, this sub-species is larger in body size and paler in color. Light hair tips are very short or absent, making the whole body color evenly pale gray. The tail and forearms are darker gray. This sub-species' tail is very long, between 72.0 and 83.8 cm (n=6), compared with 56.0 to 75.1 cm (n=13) for *T. cristatus cristatus*. Although the two sub-species have much the same head plus body length, the crown of this sub-species is dark surrounded by a zone of long hair tips (Groves, 2001).

Geographic Range

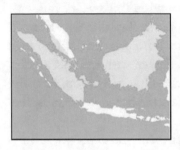 In Indonesia, *T. cristatus cristatus* is found in Sumatra, Belitung, Bangka, and on some islands in the Riau Archipelago such as Linga, Bintang, Sugi, Jombol and Bakang (Groves, 2001). But their distribution also reaches Borneo and the western coast of peninsular Malaysia. *T. cristatus vigilans* is restricted to Serasan Island, off West Borneo (IUCN, 2015).

Behavior and Ecology

The Silvered Langur is primarily a folivorous species, but also consumes fruit, seeds, flowers, and young shoots (Bernstein, 1968; Furuya, 1961). Young leaves and leaf buds are selected from the leaves of trees and leaves of any age are selected from ground plants and vines (Bernstein, 1968). Orchid flowers and dried wood have been seen to be eaten by the silvered langur (Bernstein, 1968). This species prefers to eat immature leaves over more mature ones because they contain less lignin and tannins. Clay is eaten by scraping the teeth along a clay outcrop (Bernstein, 1968).

This species is diurnal and moves arboreally using a semi-brachiator movement mode. They are fine swimmers and spectacular jumpers. The group will start moving shortly before sunrise (Bernstein, 1968). They may also travel right before it becomes dark, with all group members retreating to a single tree to sleep at night (Bernstein, 1968). When resting juveniles and infants stay close to adult females (Furuya, 1961).

Silvered Langurs live in unimale groups with several adult females and immature individuals. The average group size is 32 individuals. Males disperse from their natal group. They will live solitarily before finding a group of their own (Bennett and Davies, 1994). Groups range between 200 and 500 m each day with a home range of between 20 and 50 ha. They are territorial and actively defended their territories. Conflicts will usually occur at the areas of home range overlap. Conflicts usually consist of chasing and/or fighting between the resident males (Bernstein, 1968). Adult males will move rapidly amongst the group and emit loud vocalizations when two neighboring groups come together (Bernstein, 1968). Other members of the group will squeal and embrace each other during group conflicts (Bernstein, 1968). Fighting consists of slapping and pulling with some biting (Bernstein, 1968). Intragroup agonistic behavior is rare in the silvered langur (Bernstein, 1968).

Infanticide has been reported to occur in this species (Wolf and Fleagle, 1977; Wolf, 1980). This happens when an adult male invades the group, killing the resident adult male and killing the infants so that the adult females will begin to ovulate again (Wolf and Fleagle, 1977). An adult male that takes over a group will also chase out any immature males (Wolf and Fleagle, 1977). Females allow other females to carry and to care for their young (known as allomothering) (Fleagle, 1988). Adult males have also been observed to carry and care for immature group members. Usually when adult males carry young, the young initiate the behavior and the adult males are passive. Females will carry infants ventrally. Infants or juveniles that fall to the ground are retrieved by juveniles or adult females (Bernstein, 1968).

Social play in the silvered langur mostly occurs amongst juveniles of both sexes and infants, but individuals of all ages and both sexes have been found to participate. Wrestling is the most common form of play with locomotor play also found. Locomotor play includes repeated circuits along a pathway where individuals will swing from branches or drop to the ground from branches. Juvenile males engage in the most vigorous forms of play fighting, and they will sometimes play fight with sub-adult males (Bernstein, 1968).

Conservation Status

This species is listed on CITES Appendix II and as Near threatened on the IUCN Redlist. They are known to occur in at least four protected areas in Indonesia, Bukit Barisan National Park, Gunung Leuser NP, Kerinci Seblat NP and Tanjung Puting NP. They are threatened by habitat loss, especially due to land cleareance and forest fires which are common in their habitat range.

Where to See It

Their natural habitat is in riversides in Gunung Leuser National Park, Kerinci Seblat NP, Way Kambas and some other national parks in Sumatra and Kalimantan. Since this species is distributed widely, it can be found in many forested areas and even in rubber plantations in Sumatra and Kalimantan. The closest place to see it from Jakarta is in Way Kambas National Park in Lampung Province, Sumatra. You can fly to Bandar Lampung, which will take 30 minutes or you can drive and take a ferry to Bandar Lampung, which will take approximately 4 hours. From Bandar Lampung you need to drive 2 hours to Way Kambas National Park.

47. West Javan Langur

Trachypithecus mauritius

Other common names: Lutung (Sunda), Lutung hitam (Indonesia)

Identification

The pelage is glossy black with a slightly brownish tinge underneath, on sideburns and on legs. Compared to *T. auratus*, this species is lacking light-tipped body hairs.

Geographic Range

Their distribution is restricted to the western parts of Java (Groves, 2001).

Behavior and Ecology

This species is active during the day and arboreal, though sometimes descends to the ground. They consume young leaves, fruit, and flowers. Particular parts of food items are eaten e.g. the seeds of some fruits, the calyx of some flowers and the petioles of some young leaves. These parts of plants tend to be lower in fiber content and are more digestible than the parts that are not eaten. Phenolics and condensed tannins are not a main determinant in food selection.

They are polygamous with groups consisting of one or two or more adult males with several adult females and their offspring. Group size varies between 6 and 21 individuals with a mean group size of 10 and a home range size between 5 and 8.5 ha in secondary forest.

Conservation Status

They are listed as Vulnerable in the IUCN Redlist

Where to See It

This species can be found in several forests in West Java and Banten, such as Gunung Gede Pangrango National Park, Gunung Halimun Salak NP, and Ujung Kulon NP. The most visited place to see it is Pananjung Nature Preserve in Pangandaran. The trip from Jakarta to Pananjung Nature Preserve can take around 6 to 7 hours but with the airline will take an hour or less. You can easily find it in Gunung Gede National Park, it is only 2 hours from Jakarta, approximately 70 km to the south.

48. Proboscis monkey

Nasalis larvatus

Other common names: Bekantan (Indonesia); Kahau (Kalimantan); Bakara, Bengkara, Bengkada (Nagju, Kutai); Paikah (Manyaan); Rasong (Sea Dayak); Batangan (Pontianak); Monyet Belanda (South Kalimantan).

Identification

The body weight of an adult male proboscis monkey is between 16 and 22 kilograms, and of females between 7 and 12 kg. They display sexual dimorphism, especially in body size (Kern, 1964). The sexes are hard to distinguish until they reach adulthood and the males outgrow the females and develop their characteristic nose. The young have an upturned nose. The males have a large pendulous nose, which is smaller and more upturned in females. This nose of the male assists in enhancing vocalizations, acting as an organ of resonance (Ankel-Simons, 2000). The proboscis monkey has interdigital webbing on the hind feet, a physical characteristic shared with no other primate, which may be an adaptation for swimming. While the function of the large nose has yet to be definitively explained, the feet confirm this species to be the most aquatic of all primates.

The pelage color is pink and brown with red around the crown, nape, and shoulders, and with gray on the arms, legs, and tail (Kern, 1964). There are cream-colored patches on the cheeks and throat. The males have a black colored scrotum and a penis that is red in color (Ankel-Simons, 2000). When the infants are born their faces are a vivid blue (Pournelle, 1967). At age 2.5 months the facial color of the infant darkens to a sooty dark gray, and then at 8.5 months the gray starts to lighten to the flesh color of the adult (Pournelle, 1967).

Geographic Range

The proboscis monkey is found on the island of Borneo. In general, it prefers to live in nipa-mangrove-mixed forest, mangrove forest, and lowland forest. In Sarawak it was found to prefer riverine and mixed diterocarp/high kerangas forests (Salter et al., 1985). It has been suggested that this species is restricted to the coastal areas and areas nears rivers because the interior has soils that are low in minerals and salts, which are needed in the diet (Bennett and Sebastian, 1988). It avoids areas with heavy deforestation, such as agricultural land (Salter et al., 1985).

Behavior and Ecology

This species is diurnal, mostly arboreal, a semi brachiator, a spectacular jumper and a fine swimmer. It is included with the colobine monkeys because it possesses a fermentative digestive system. As a result, the diet is restricted to vegetative matter with no glucose content. They therefore only eat the non-sweet parts of fruit, flowers, young leaves and pedicels. Proboscis monkeys are very selective feeders and a group will move directly between food sources. Groups are led by a single dominant male. Young and sub-adult males are excluded from the harems and form all-male bachelor groups. These bachelor groups follow established harems closely but are more mobile, often travelling many kilometers in a day. Harems also frequently travel together, forming large associations of more than one harem but are always discernable as different groups by the dominant males who tend to maintain their distances from each other.

These groups may travel for many days together. Groups frequently keep in contact with other groups by calling. The adult males make distinct loud, far carrying resonant calls, intermittently throughout the night. They sleep along the river's edge at night, which increases visibility serving an important need for the group to remain in contact, which suggests in turn that these monkeys require both vocal and visual contact. The proboscis monkey is a well-known swimmer; it may use water as an escape from arboreal predators such as the Clouded Leopard, probably the only forest carnivore in Kalimantan able to take a proboscis monkey. Crocodiles may be other predators, because proboscis is often swimming across the rivers.

The most important plant species used by this species in Tanjung Puting National Park were *Eugenia sp., Ganua motleyana*, and *Lophopetalum javanicum* (Yeager, 1989). At Gunung Palung

Nature Reserve, the most important plant species consumed were *Mesua lepidota, Palaquium sp., Baccaurea lanceolata, Barringtonia racemosa,* and *Salacia macrophylla* (Ruhiyat, 1986). Also, at Gunung Palung Nature Reserve the most frequent tree species visited for fruit were *Syzigium borneense* and *Dialium indicum* (Ruhiyat, 1986). At Samunsam Wildlife Sanctuary in Sarawak the majority of foods consumed were found to be fruit or seeds, with young leaves the next most important food item (Bennett and Sebastian, 1988). Fruit eaten tend to be of the dry and bitter-tasting type (Bennett and Sebastian, 1988). In Sarawak it was found that the food item representing most species, were leaves (Salter et al., 1985).

Group sizes range from 3 to 32 individuals. This is not a territorial species with group ranges overlapping (Kawabe and Mano, 1972; Boonratana, 2000). Groups of proboscis monkeys will usually never move father than 600 m from a river or stream (Bennett and Sebastian, 1988). They are most active from late afternoon to dark (Ruhiyat, 1986). Groups will move away from the riverine forest in the morning and into the mangrove forest in the afternoon, and back to the riverine forest in the evening (Salter et al., 1985; Boonratana, 2000). This species has a high dependency on habitats that adjoin rivers (Boonratana, 2000). At Sukau, in Northern Borneo, was found to travel at a mean height of 12.25 m, and at Abai the mean height for travelling was found to be 6.91 m (Boonratana, 2000).

The proboscis monkey tends to cross rivers and streams at narrower points which could be an anti-predator behavior especially avoiding crocodiles (Yeager, 1991); males tend to be the last individuals to cross amongst group members (Yeager, 1990). Also this species will cross rivers in large numbers, use foliage to "spring-board" across rivers, and visually scan searching for predators (Yeager, 1991). When individuals cross a river alone, they enter the water quietly and swim to the other side, exiting quickly and hurrying up a tree (Yeager, 1991). The false gavial, *Tomistoma schlegeli* is a major predator of this species, and the river crossing behavior may be related to avoiding this and other riverine predators (Galdikas, 1985).

Conservation Status

They are listed under CITES Appendix I and classified as Endangered in the IUCN Red list.

Where to See It

Kalimantan or Borneo island generally, including Pulau Kaget and Pulau Laut in South Kalimantan are home to proboscis monkeys, which are limited to the lowlands. They are most commonly seen on river banks, and in swamps, inland swamps and mangrove habitats. In Kalimantan, the species is well represented in Gunung Palung in western Kalimantan, where coastal mangroves grade into riverine forest eventually rising to dry lowland and then hill forest; Tanjung Puting National Park in Central Kalimantan, a logged peat swamp forest with areas of freshwater swamp forest; Mahakam river basin; Kutai National Park, which also has peat swamp forests in eastern Kalimantan; and Danau Sentarum Wildlfe Reserve in West Kalimantan. In Tanjung Puting, you can see this species sitting in trees along the rivers. They can be seen in the morning and late afternoon before feeding in the forest interior.

49. Pagai Pig-tailed Langur

Simias concolor

Other common names: Simakobu (Siberut), Masepsep (Sipora, North Pagai, South Pagai), Masesep simabulau (North Pagai).

Identification

Body length of the Pagai Pig-tailed langur ranges from 49 to 55 cm in males and from 46 to 55 cm in females. Males are slightly heavier than females with the average weight of males at 8.7 kg and 7.1 kg for females. Body color varies for Simakobu: black and creamy, with white cheek patches. The face color is black with a small snub-nose. There are two subspecies found in Indonesia, *Simias concolor concolor* and *Simias concolor siberu* (Groves, 2001).

Geographic Range

Simakobu is endemic to Indonesia and confined to the Mentawai islands off the western coast of Sumatra. *Simias concolor concolor* is distributed throughout the Mentawai islands: Sipora, North Pagai, South Pagai and some offshore islets. *Simias concolor siberu* is only found on Siberut Island (Groves, 2001; IUCN, 2014)

Behavior and Ecology

This species is active during the day and moves both arboreally and terrestrially. They mainly consume leaves with some fruit, seeds and flowers. Occasionally, they also raid cacao plantations, making them pests. They live in groups consisting of 5 to 15 individuals with one adult male (Whitakker, 2005).

They are known to be easy to hunt. When they feel threatened, they tend to stay motionless in dense foliage and not make a sound. If the threat continues, they drop to the ground and attempt to flee. Simakobu rarely call and they usually move slowly and silently.

Hunters usually catch them using poisoned arrows. Based on field research, a male simakobu who loses his female will seclude him self for 20 days (Whitakker, 2005).

Conservation Status

The main threats to this species are heavy hunting, both as game and for the pet trade and habitat destruction. Whittaker (2006) observed that Simakobu hunting was formerly regulated by local taboos and rituals, but the penetration of Christianity into the area has changed local wisdom. Furthermore, the threat has increased since a shift by hunters from bows and arrows to rifles. Habitat destruction due to commercial logging, establishment of oil palm plantations, and general forest clearing and product extraction and the accompanying road construction into remote areas also have an impact on the population of this species. Based on IUCN Redlist, simakobu is listed as Critically Endangered on the IUCN Redlist and appears in Appendix I CITES.

Where to See It

This species is rare and shy so it can be a little bit tricky to find them. They usually hide high up in trees. Siberut National Park is the ideal place if you want to look for simakobu. To reach Siberut Island you have to cross the ocean, which can be done with a rental boat from Padang, West Sumatra. There is also a ferry from Padang Harboar to Siberut Island. The journey can take from 8 to 10 hours with slow boats but currently there are some faster boats that will take 4 hours.

D. FAMILY HYLOBATIDAE (Gray, 1870)

Gibbons and Siamangs are medium sized primates (3.9 to 12.7 kg) of the tropical rainforests of South-East Asia, North-East India, Southern China and Bangladesh. They lack the tail of other arboreal apes. They have remarkably long hands and feet with a deep cleft between the first and second digit of their hands. They entirely live in trees and are known as the most agile forest acrobatic. Their mode of locomotion is brachiating; that is by hanging and then swinging from one branch to another. Movement through the tree tops is very fast and can reach 60 km/hour.

Gibbons and Siamangs are monogamous. Pairs live together, usually with a maximum of two offspring. The pregnancy period is 7 to 8 months; inter-birth range is 2 to 2.5 years. Sexual maturity is between 8 and 9 years. Life span is 25 years. They are territorial and maintain their territories with conspicuous vocalization. They have loud, back and forth voices. In the morning gibbons and siamang always produce a shrieking morning call that can be heard for a very long way. The male usually calls first and then the female responds. Each species has a different rhythm and voice character, which can be seen in a sonography. Some experts use sound for their field identifications.

Hylobates: Gibbons

The genus *Hylobates* was previously considered to be the only genus in the family Hylobatidae, but recently its subgenera *Hoolock* (formerly *Bunopithecus*), *Nomascus*, and *Symphalangus* were elevated to the genus level. The genus *Hylobates* contains eleven species in the world, eight of which occupy Indonesian rain forests. *Hylobates agilis* found

J. Supriatna, *Field Guide to the Primates of Indonesia*,
https://doi.org/10.1007/978-3-030-83206-3_7

in Sumatra, but also occupy Peninsular Malaya. While *Hylobates klossii*, which is predicted as the oldest member of Hylobatidae family, is an endemic primate in Mentawai islands. *Hylobates lar vestitus* (one of the subspecies of *Hylobates lar*) is native to Sumatra island. The species of *Hylobates* found in Kalimantan rain forests are *Hylobates albibarbis*, *Hylobates muelleri*, *Hylobates abbotti*, and *Hylobates funereus*.

Symphalangus: Siamangs

Siamangs belong to what was formerly a subspecies of *Hylobates* that has recently been elevated to the genus level called *Symphalangus*. Some authorities recognize two distinct subspecies; the mainland Asia and Sumatran population (*S. syndactylus continentis* and *S. s. syndactylus*) (Mittermeier et al., 2013). But no subspecies are recognized by Groves (2005). Within the family of Hylobatidae, siamangs are the heaviest up to 13 kg. A group of Siamangs usually produce back and forth sounds, which can be heard kilometers away in the forest. Siamangs are sympatric with *Hylobates lar* in northern Sumatra and *H. agilis* in southern Sumatra.

Figure 9. Indonesian gibbons: a. *Hylobates agilis* (Sri Mariati); b. *Hylobates albibarbis* (Noel Rowe); c. *Hylobates klossii* (Gusmardi Indra); d. *Hylobates moloch* (Nurul L. Winarni & Anton Ario); e. *Symphalangus syndactylus* (WCS-IP); f. *Hylobates funereus* (Indra Hana); g. *Hylobates lar* (Noel Rowe); h. *Hylobates abotti* (M. Khotiem); i. *Hylobates muelleri* (Noel Rowe)

50. Agile Gibbon/ Black-handed Gibbon

Hylobates agilis

Other common names: Ungko lengan hitam (Sumatra); Kelawat (southern Sumatra);

Identification

Previously, two subspecies were recognized, *Hylobates agilis agilis* and *Hylobates agilis ungko* but this is currently in debate and there now seems to be a general consensus that this species is monotypic (Groves, 2001). *Hylobates agilis* adults weight between 5.0 and 6.4 kg with a body length between 45 and 50 cm. They have no tail. They have white eyebrows and there is a white cheek pathe in the males. Generally, body color varies among three phenotypes from buff to black. The brown phenotype of this species is nearly the same as the browner part of the *H. albibarbis* spectrum but lacks black on the hands and feet. The buff phenotype is pale, with browner under parts. The black phenotype is glossy black all over the body except for a dark brown lower back.

Geographic Range

Hylobates agilis is found southeast of Lake Toba and the Singkil River on the island of Sumatra. They are also found in Peninsular Malaysia from the Mudah and Thepha Rivers in the north to the Perak and Kelantan Rivers in the south and in South Thailand near the Malaysian border, east of the Thepha River watershed (Gittins, 1978; Groves 2001; Marshall and Sugardjito, 1986).

Behavior and Ecology

Agile gibbons are frugivorous. Their main diet consists of fruit, including large quantities of figs (*Ficus spp.*), leaves, flowers, and insects. Fruit comprises about 60% of the diet, mostly ripe, sugar rich, juicy fruits, and leaves 30%, but monthly proportions of fruit sometimes exceed 90% or fall below 30%. They are diurnal and completely arboreal. In the jungle, they move by leaping from branch to branch. They are spectacular acrobats but sometimes they are seen walking bipedally on the ground or on big branches. They are commonly seen in all forests up to 1,000 m above sea level, excluding mangroves. They are normally found only in primary and selectively logged dipterocarp forests.

Agile gibbons are monogamous and territorial. They have a small territory of 29 ha., which is defended. They travel about 1,200 m per day and often take only 10 to 15 minutes to cross from one side of their territory to the other. They normally take 2 to 3 days to cover their ranges completely. There is no strong evidence for patrolling and boundary disputes normally do not affect the ranging pattern. On average there are 4.3 groups/km² with group size averaging 4.4 individuals.

They defend their territories with regular loud morning calls. Normally found Low levels of interactions are normally found in gibbons due to a lack of social partners. This is particularly the case for older infants and juveniles. Compared to the young of most other primates, they have no sameaged playmates because of their small group sizes. Female Agile gibbons are often co-dominant with male in their group. This is in contrast to many other species of primates. Intergroup encounters occur only once every two days and conflicts with other groups are infrequent. Fighting is rare. Female participation in border disputes is normally limited to calling. Sub-adult and adolescent males often participate in intergroup conflicts and join their parents in chasing a neighboring male.

Dawn singing of male agile gibbons begins simply, two or three notes at a time. During the half hour or more of the ensuing chorus, short songs of up to 14 notes develop that include a diphasic couplet. Equally diagnostic is the ascent in the pitch of the opening short notes of the song, the first being lowest with only one exception, a song that begins high, apparently with an inhaled "hah". Consistent intervals are maintained, varying from two to five songs per minute according to the individual. The male agile gibbon avoids double inflections. The female agile gibbon indulges in several minutes of monotonous prelude (warmup) and puts a flourish on the start of the first note of the great call. The ensuing great call, beginning with the first long note, is of the soaring type as opposed to trilling. Great calls of agile gibbons differ from one female to another in the number of notes and duration.

The solos of unmated males tend to be longer and temporally separated from those of mated males, and apparently function to attract unmated females. The female's song is sung as a duet with the male. Solitary, transient females almost never call, and widowed females tend to sing irregularly if at all. The female's loud song, therefore, seems to advertise her paired status as well as her presence on a territory. Duets can occur anytime between dawn and ten o'clock in the morning. Duets begin spontaneously or in response to a call by a neighboring group. They last about 15 minutes.

Conservation Status

Loss of habitat due to forest clearing for agricultural and logging has led to a declining population of this species in the wild. This species has lost 66% of its habitat, from 500.000 km^2 down to 170.000 km^2. It is estimated that there are only 30,000 individuals left in the wild. At this moment, It occurs in protected areas only in Kalimantan and Sumatera. This agile gibbon is listed as an endangered species on the IUCN Red list and is likely to become extinct if there is no immediate action. The Indonesian government has a law to protect the gibbon.

Where to See It

They can be seen in several National Parks across Sumatra Island such as Way Kambas NP, Barisan Selatan NP, Kerinci Seblat NP, and Gunung Leuser NP. The easiest access is in Way Kambas NP. This lowland forest has more than 100,000 ha and harbors other charismatic wildlife such as Elephants, Rhinoceroses, Tigers, many other different cats, monkeys and birds. It's located in Western Lampung province, Sumatra. From Jakarta you can fly to Bandar Lampung or drive via the Java-Sumatra highway across the Sunda strait. From Bandar Lampung the drive to the park will take approximately 3 hours. Hotel and basic needs are available close to the park and return trips to the main city in the province, Bandar Lampung, are also possible.

51. Muller's or Bornean Gibbon

Hylobates muelleri

Other common names: Wau wau (South Kalimantan)

Identification

Muller's gibbon has a white brow band. It is greybrown in color but with a wide range in coat color and pattern. It is paler on the rump than on the back and black on the throat and axilla (armpit) if not the entire abdomen. Adult Muller's gibbons weight between 5.0 and 6.4 kg. They can be detected by the loud, bugling call of the adult female.

Geographic Range

This species is found in southeast Kalimantan, approximately south of the Mahakam River and west of the Barito River

161

Behavior and Ecology

This gibbon is active during the day and moves arboreally. A study of this species in Barito Ulu revealed that this species consumes a large amount of non-fig fruits, followed by young leaves, figs, and flowers. It also consumes a small number of insects Fruit is available throughout the year but peaks in abundance and diversity from March to May. Muller's gibbon tends to prefer medium size, yellow-orange fruit with juicy pulp and with a thin or rind-like skin, in dense but small crops. They tend to avoid purple-black and red fruit with soft dry pulp, many-seeded fruit in sparse crops or fruit with large seed (more than 20 mm wide).

As with other gibbons, they are monogamous and territorial and their territories are defended with regular loud morning songs. They live in family groups composed of two to six members (adult male, adult female, juveniles, and infants). The adult male is the principal protector of the group and the more active aggressor in intergroup conflicts. The population density in Sungai Wein protection forest is 7.9-9.5 ind/km2 (Mittermeier et al., 2013). But in many places, group density may be 3 or more groups/km^2 with a mean group size of 3.4 individuals.

Males sing almost every morning from their sleeping trees before dawn. Most songs begin high and descend. Male evidently listen to their neighbors between songs, then usually make response call. Females usually initiated duets, while males will follow. The loud call of females can be heard for a distance of over 2 km under suitable conditions. Song duration of this gibbon also increases with intruder pressure.

Conservation Status

Habitat destruction and hunting for the wildlife trade and for human consumption pose a threat to the survival of Muller's Gibbon. The population is estimated to have been reduced by over 50% in the past 45 years (3 generations). Based on that reason, *Hylobates muelleri* is listed as Endangered on the IUCN Red list.

Where to See It

Normally they occur primarily in the lowlands and hills of southeastern Kalimantan, such as Barito Ulu in Central Kalimantan province and Meratus Protected Area in South Kalimantan province. They have not been found in mangrove forests. To see Meratus forest, from Banjarmasin, the capital city of South Kalimantan province, you need to rent a car to drive to Loksado of Hulu Sungai Regency for about 165 km, the foothill of the Meratus mountain (up to 2000 asl). The mountains are home to the "Orang Gunung" or Mountain People, an ethnic group collectively referred to as the mountain Dayak. In times past the Dayaks lived a semi-nomadic lifestyle in small longhouse communities scattered throughout the range. Home stays are available in Loksado in the villages scattered along the many mountain trails.

52. Kloss's Gibbon

Hylobates klossii

Other common names: Siamang Kerdil (Indonesian), Bilou (Mentawai)

Identification

Adult *Hylobates klossii* weigh around 5.5 kg with both male and female body length around 45 cm. They have no tail. They are totally black without any individual differences in pelage color or pattern, which aids the identification of individuals in other species.

Differences between the sexes are minimal. Male genital tufts are present in all gibbon species except *H. klossii*. The labia majora are prominent in gibbons, and the labia could sometimes be mistaken in the field for the small penis of the male. The females urinate straight down, whereas male urine is directed backwards. Female nipples are larger than male nipples but the difference is only relative and is not a very useful character. It is difficult to differentiate males and females in the wild.

Geographic Range

They are found only on the larger islands of the Mentawai Islands group, Siberut, Sipora, North Pagai, and South Pagai, which lie 85 to 135 km off the west coast of Sumatra, Indonesia.

Behavior and Ecology

The diet is mainly fruit, 80%, and leaves and insects, 20%. Kloss' gibbon is an animal that requires fruit trees that ripen during different times of year in order to provide a regular food source. This gibbon also eats the leaves of lianas, leaf shoots of orchids, and bark of trees such as *Shorea spp.*, *Hopea sp.*, and other Dipterocarps.

Just like other species of gibbons, Kloss's gibbon is diurnal (active during the day) and arboreal (live on the tree). They use both of their arms to move from branch to branch. They may also be found walking bipedally on the forest floor, although this is rare. Their daily movement ranges across 1 - 2 km, and their territories cover 7-32 ha, which they always defend. They defend their territory using calls as well as physical confrontation.

Their territorial great call is refined, simple, slowly modulated, and flute like. Males sing mainly from their sleeping trees in a predawn chorus of short songs gradually elaborated to include a trill. Conversely, females sing later in the morning. The early morning great call of males is uttered at 1minute intervals. Each great call lasts 20 sec or more with soft opening notes that swell in volume, reaching a climax in pitch, intensity or rapidity, and then subsiding. The pitch varies between 0.5 and 1.5 kHz with few exceptions. The song starts as an ascending series of sustained notes that become shorter as they gain a terminal upward inflection. Females in adjacent territories often meet at a shared territorial boundary and sing together. Daughters still with their parents sing with their mothers, but sons do not sing with their fathers.

In physical confrontations, males defend against males, females against females. Adult males are the principal protectors against predators and the more active aggressors in intergroup encounters. The adult female almost always travels first in group progressions.

Similar to other species of gibbons, Kloss's gibbon lives in monogamous pairs. Group size is 4-6 individuals. The average group size is 3.7 individuals. The couple is usually accompanied by offspring under 8 years old. Offspring that have reached 8 years, whether female or male, separate themselves from their parents.

The mothers usually help young adult offspring to find an empty place for their territory. They sleep in branches of trees that have no lianas, and rarely stay in the same tree for more than two days. This could be a strategy to avoid predators, generally pythons. In the wild, Kloss' gibbon can live side by side with other primates, such as the Lutung Mentawai *(Presbytis potenziani)*.

Conservation Status

Although Mentawai people generally hunt primates, Siamang Kerdil is not their main target. Habitat destruction has contributed more than hunting to the decline of this species. Originally, they occupied an area as wide as 6,500 km² on the Mentawai islands, but the area of habitat ha shrunk by 31% to 4,500 km. The population left is estimated to be 20,000-25,000 individuals. Only 1,490 km² of their range is protected. Because it is endemic, this primate has been protected under the law of the wildlife protection regulation 1931 no. 266, the ministry of forestry decision 10 June 1991 no. 301/Kpts-II/1991 and UU no 5 1990. The species is listed as Endangered on the IUCN Red list.

Where to See It

They are most likely to be seen in primary forest, but sometimes also in nearby logged areas. They do not occur in mangrove and coastal forests. Kloss's Gibbon can be found in Siberut NP on Siberut Island. They have a distinct call and can be used as guidance to find them. Access to Siberut:

- Padang ──▶ Muara Siberut/Muara Sikabaluan on Siberut Island is approximately 155 km and takes about 10 hours on a ferry boat, which operates three times a week. Siberut NP can be reached from Muara Siberut/Muara Sikabaluan using a motor bike.

    On Siberut, accommodation is available in the Syahruddin Hotel and the Wisma Tamu Guesthouse. Local guides also avalable in Padang, Bukittinggi and Muara Siberut. To find local guides you can check in Jl. Pontianak N/13 Padang, West Sumatra.

53. Sumatran Lar Gibbon/Sumatran White-handed Gibbon

Hylobates lar

Other common names: Ungko Lengan Putih Sumatra (Sumatra)

Identification

There are five subspecies of *Hylobates lar, Hylobates lar vestitus, H. l. lar, H. l. carpenteri, H. l. entelloides,* and *H. l.*

yunnanensis, but only the subspecies *Hylobates lar vestitus* occurs in Indonesia. These subspecies have well-defined latitudinal distributions, but in general, they are not highly distinct. The differences are based largely on relatively minor body color variation and the degree to which the fur is multi-colored (Brockelman, 1985, 2004; Groves, 2001; Woodruff, 2005). Body color varies from white brownish to black or black brown to red. Besides covering the arms and feet, the white hair on this species also grows around the face. Males and females show the same color.

The length of the female body is between 42 and 58 cm, and the male is between 43.5 and 58.5 cm. Female body weight is between 4.5 and 6.8 kg, and male body weight is between 4.9 and 7.6 kg. Males and females have the same canine teeth.

Geographic Range

Hylobates lar vestitus occupies Northern Sumatra, Northwest of Lake Toba and the Singkil River. The other four subspecies are found throughout Peninsular Malaysia, (except for a narrow strip between the Perak and Mudah Rivers, where *H. agilis* occurs), north through southern and eastern Myanmar east of the Salween River, throughout most of Thailand, though not in the north-east, and marginally into southern China.

Behavior and Ecology

Sumatran Lar gibbon diet consists of ripe fruit, young leaves, old leaves, some flowers, buds, and insects. Monthly proportions of different foods consumed are fruit, mean 50%, (range 36 to 60%), 22% of which is figs, leaves 29% (14 to 53%), flowers 7% (3 to 8%), and insects 13% (6 to 24%). In Ketambe, North Sumatra, which has a figrich forest, Siamangs and Lar gibbons Sumatran White-handed gibbons with overlapping ranges, have foods more similar to each other than to neighboring groups of the same species.

As with other gibbons, they are territorial and monogamous with families consisting of two to six members, and territories, which are defended through loud song in the morning. In one the male was dominant over the female. In the other dominance varied. In late pregnancy and early postpartum, the female became dominant over the male in feeding behavior. The adult male is the principal protector from predation (as in *H. klossii* and *H. muelleri*) and the more active aggressor in intergroup encounters.

The males song at dawn are normally continuous. Sometimes discrete songs are uttered in a dawn chorus or at midday, when two males may duel vocally. Males are unable to be distinguished from one another by their songs. The female great call resembles that of the agile gibbon. The song of Lar gibbons resembles others such as Kloss' gibbon, and Siamangs and isless frequent on rainy or windy days and is positively correlated with fruit abundance and mating activity.

Intergroup conflicts are infrequent. Fighting is rare. There is consistently less evidence of fighting injures among females than among males. Sub-adult and adolescent males also participate in intergroup conflicts. A sub-adult male gradually took over the male role in territory defense from aging fathers.

Lar gibbons are active from morning to afternoon and spend most of their time in trees. The male does not tolerate the present of other males in its territory. During day rest, males and females groom each other. Lar gibbons can live side by side with Siamangs and Agile gibbons. But when they encounter long tail macaques (*M. fascicularis*) confrontation by exposing sharp fangs occur, especially over food. This species sleeps on branches, and avoid trees with lianas for predator avoidance. They also rarely sleep in the same tree for more than two days. This could be to avoid parasites.

This species is completely arboreal and diurnal. The home range is variable, from 44 to 54 ha in West Malaysia and 16 ha in Thailand, but there is no information on the size of the home range in North Sumatra. They travel an average of 1.5 km per day. Group density is 2 groups/km^2, with an average group size of 3.3 individuals. They occur in all forests up to 1,000 m above sea level, excluding mangrove and beach forests.

Conservation Status

With its habitat shrinking by 55%, from 68,000 km^2 to approximately 30,880 km^2, the Sumatran Lar gibbon has become vulnerable to extinction. It is estimated that the current population size is only 250,000 individuals and they are found in only 4000 km^2 of protected areas. *H. lar vestitus* has been protected in Indonesia according to the wildlife protection regulation no 299, 1931, the ministry of forestry letter of decision 10 June 1991 No. 301/Kpts-II/1991, and Law No. 5 year 1990. *H. lar vestitus* is listed as Endangered on the IUCN Red list.

Where to See It

The best place to see this primate is in Gunung Leuser National Park, whether in the North Sumatra or Aceh parts of the park. It can also be seen in Janto Nature reserve in North Aceh. Gunung Leuser NP covers more than 1 million ha, but go to Ketambe Research Station, located close to the city of Kutacane, approximately 4 to 5 hours drive from Medan, North Sumatra. Gunung Leuser NP has several entry points which are Lawe Gurah, Ketambe, Bohorok – Bukit Lawang, and Sikundur – Besitang. From Medan Gunung Leuser NP can be reached through these routes:

- Medan ⟶ Kutacane ⟶ Lawe Gurah/ Ketambe (± 275 km). using public transport (bus or taxi) this will take 6 to 7 hours. Buses depart the Pinang Baris bus terminal in Medan to Kutacane around 15 times per day, while from Kutacane to Lawe Gurah/ Ketambe, buses depart twice a day. Lawe Gurah is a Tourist attraction 43 km from Kutacane. In Kutacane there is an orangutan research station.
- Medan ⟶ Bohorok/ Bukit Lawang (± 91 km) can be reached by bus in 2.5 hours. In Bohorok there is an orang utan rehabilitation station and there is accommodation in Bukit Lawang.

If you are in Banda Aceh, after a one hour drive you can see this animal in Janto protected area or at Cut Nyak Din Grand Forest Reserve.

54. Javan Gibbon or Silvery gibbon

Hylobates moloch

Other names: Owa, Wau-wau Kelabu (Jawa Barat)

Identification

Research by Andayani et al., (2001) suggested that *H. moloch* has two genetically distinct populations, which lead to the recognition of two sub-species, *H. moloch moloch* and *H. moloch pangolsini*. Geissmann et al., (2012) cast doubt on this claim and the Silvery

Javan gibbon is currently recognized as a monotypic species, using molecular evidence and a comparison of their morphological and vocal data characteristics.

Male Javan gibbons weigh between 4.3 and 7.9 kg while females weigh between 4.1 and 6.8 kg. In general, body coloration of this species varies from blackish to silvery grey to a grey-brownish color. The face is totally black with pale eyebrows. There is a slightly paler hair color surrounding the face compared to the rest of the pelage and a dark crown on their breast hair. They have either a black or grey cap and black genital spots are present (Geissmann, 2002, Ario et al., 2018).

Geographic Range

The Javan gibbon is endemic to the island of Java and is mostly confined to the western part of that island (Banten and West Java). They are also known from Central Java as far east as the Dieng Mountains. In western Java, they occur in areas that still have original rainforest cover, which is the most suitable habitat. The distribution of Javan Gibbon can be determined based on their elevation preference. They are commonly still found between 1,400 and 1,600 m above sea level and have never been seen above 1,600 m. This is presumably because the vegetation composition changes at higher elevations (Supriatna et al., 2010, Setiawan et al., 2012).

Behavior and Ecology

Food sources consist of fruit (61%), leaves (38%), flowers (1%), and insects. They are known to feed on 125 different plant species. These food resources are usually found within their home range, which averages 17 ha and is well defended by a family. They usually forage in the morning and evening, while in the afternoon they are usually seen resting or grooming.

Just like other specie of gibbon, Javan gibbons live in monogamous family groups. A group usually consists of one adult male and one female with 1 to 2 dependent offspring. Generally each family occupies their own territory. They average 1,400 m of travel during the day. They are arboreal, which means they are dependent on forest with a closed canopy. They are very rarely seen on the ground (Supriatna et al., 2010).

Female Silvery Javan gibbons call in a duet with the male. Females also call alone or in response to male solos. Apparently, males approach intruding pairs alone while their mates sing. Consistent individual differences easily distinguish neighboring female Silvery Javan gibbons, thus compensating for the lack of a family labeling male song. Female participation in border disputes is commonly limited to calling out an extreme scream, or great call, throughout an encounter, during which the female may become more aggressive. Morning calls are only made by adult females.

The morning call behavior by adult females is an effort to communicate with other groups and express territoriality (Kappeler 1981). Vocalization in Javan gibbon is unique compared to that of other Hylobatids, in that females play a greater role in keeping territory (Geissmann et al., 2005). Alarm calls can be produced by either adult males or females, but not by sub-adults, juveniles, or infants.

Conservation Status

The increasing human population in Java has caused the tropical rainforest to decline drastically. Originally the Javan gibbon occupied an area of more than 40 thousand km^2, but today only 4%, or about 1, 600 km^2 is left. Besides that, they are hunted for pets. As a result, almost all remaining individuals are restricted to several national parks (Gunung Gede Pangrango NP, Gunung Halimun Salak NP, Ujung Kulon NP, and possibly in the Gunung Ciremai NP, plus several small protected areas in West Java and Central Java (Ario et al., 2018).

Silvery javan gibbons are listed as Endangered on the IUCN Redlist. In 1987, the IUCN Species Survival Commission considered them to merit the highest conservation priority of all Asian primates. No large section of the population is really secure. Despite the 1931 wildlife protection regulation, strengthened with the Ministry of Forestry letter of decision 10 June 1991 No. 301/Kpts-II/1991 and Law (UU) no 5, 1990, the population in the wild is still declining. At this point, it is predicted that there are only 2,000 to 4,000 left in the wild. They are recognized as one of the Indonesian primates closest to extinction.

Apart from ex-situ conservation efforts, habitat protection and the enforcement of current laws and regulations are really needed to save this endangered primate. Fewer than 100 wild-caught Javan gibbons are in 8 rehabilitation centers and 11 zoos on the Islands of Java and Bali (Nijman 2004). Javan gibbons in rehabilitation centers usually were pets, which were confiscated by the government or given up voluntarily by the previous owners. Rehabilitation and release of Javan gibbon is not easy. Besides being monogamous and highly selective of food and mates, Javan gibbons have a strong sense of territoriality, resulting in a lengthy and costly rehabilitation process. Rehabilitation and reintroduction programs have been widely used as an element of conservation strategies for endangered species (Kleiman 1989)

Where to See It

H. moloch is very easily seen in several National Parks in West Java such as Gunung Gede Pangrango NP, Gunung Halimun Salak

NP and Ujung Kulon NP. They can also be found in some Nature Reserves such as Gunung Simpang, Gunung Salak, Gunung Ceremai, Gunung Slamet and Gunung Prahu. The closest and easiest to access from Jakarta is Gunung Gede Pangrango NP or Gunung Halimun Salak NP. Gunung Gede Pangrango NP has a lot of visitors during the weekend. This National Park has four entry points:

- Cibodas is the main entry to the park and you can find the official park office there. It is about 100 km or around 2,5 hours to Bogor from Jakarta, then 36 km or around 1 hour from Bogor through Ciawi, It is 89 km or around 2 hours from Bandung through Cianjur and Cipanas.
- Gunung Putri. This entry is 15 km from Cibodas and can be reached through Cipanas and Pacet.
- Selabintana. This entry is 60 km from Bogor and can be reached in about 1,5 hours. It is around 90 km from Bandung, which will take about 2 hours. From the main city of Sukabumi, there is public transportation, which stops at Pondok Halimun, from this point you can rent a motor bike or hike for 6 km to the entry point of Selabintana.
- Taman Wisata Situ Gunung is about 15 km from Selabintana and can be reached using public transportation from Bogor or Sukabumi (stop at Cisaat).

Gunung Halimun Salak NP can be reach from three directions:
- By car from Bogor to Leuwiliang, which is 20 km and takes about 30 minutes on public transport ⟶ Leuwiliang to Nanggung is a further 15 km and takes about 20 minutes on public transport ⟶ Nanggung to Cisangku is another 15 km from there and can be reached with a motor bike ride of an hour.
- Rangkasbitung (Lebak) – Bayah (150 km) 2 hours on public transport ⟶ Bayah to Ciparay (36 km) using bus or motor bike ride for around 2 hours

- Sukabumi – Parungkuda (20 km) using bus for 30 mins ⟶ Parungkuda – Cipeteuy (30 km) using public transport for aroung 1 hour ⟶ Cipeteuy – Perkebunan Teh Nirmala (Citalahab) motor bike ride.

55. Bornean White-bearded Gibbon

Hylobates albibarbis

Other common names: Wau-wau atau Wau-wau Brewok Putih

Identification

The Bornean White-bearded gibbon was originally thought to be a subspecies of *H. agilis* but in 2001 Groves recognized it as a distinct species. This species hybridizes with *H. muelleri* in Central Kalimantan. Morphologically, it is similar to *H. agilis* but there are no individuals with black coloration. In general, the body is bright brown all over except that the inner side of the arms, chest, abdomen and feet are black. The head is blackwith white hair framing the face

Geographic Range

This primate is found in the southern part of Kalimantan on the island of Borneo, from the southern section of the Kapuas River in the west to the Barito River in the east (Marshall & Sugardjito 1986)

Behavior and Ecology

Similar to other gibbons, *H albibarbis* is diurnal. Their time is divided into feeding (29%), travelling (29%), resting (29%), and other social and behavioral activities (13%). This species feeds mostly on fruit especially ripe fruit with high sugar content, but they are also seen eating young leaves and insects.

The Bornean White-bearded Gibbon is usually found in primary, secondary and some selectively logged tropical evergreen forest, as well as peat swamp forest (Buckley 2004; Buckley *et al.,* 2006). This species is common and easy to see in suitable habitat. Within the mixed-swamp forest of Sabangau, the population is estimated to be 19,000 individuals, representing one of the largest remaining continuous populations of the species (Buckley, 2004; Buckley *et al.,* 2006). Density estimates range from 7.4 individuals/km^2 in the logged peat swamp forest of the Sabangau catchment, and 8.7 individuals/km^2 in the heath forest of Tanjung Puting National Park (Mather, 1992), to 14.9 individuals/km^2 in a mostly mountain forest area of the Gunung Palung Reserve (Mitani, 1990), with densities decreasing at higher elevations. Average home range sizes of 28 ha and 45 ha were observed at Gunung Palung and the Sabangau catchment area, respectively (Buckley *et al.,* 2006).

Conservation Status

This species is listed as Endangered on the IUCN Redlist.

Where to See It

Gunung Palung National Park has more information available on primates than any other national park in Kalimantan. The park collaborates with Hardvard University to study its biodiversity. There is a research center there that was built in the 1980's by Dr. Mark Leighton, where research on orangutans and forest ecology is conducted. Dr. Cheryl Knott from Boston University also works at this center and studying primate ecology and behavior. The research station is in Cabang Panti, in the western part of Gunung Palung National Park. It is open for visitors/tourists. In the surrounding villages near the beach there are options for housing, motor boat rental and tourist guides.

There are several options for reaching Gunung Palung National Park, such as:

- Pontianak ⟶ Ketapang takes 1 hour 15 minutes by plane and then land or river transportation to the park takes 6 - 10 hours.
- Pontianak ⟶ Ketapang using express motor boat takes 6 to 7 hours. Ketapang ⟶ Sukadana or Teluk melano takes a further 5 hours.
- Rasau Jaya (Pontianak) ⟶ Teluk Batang takes 4 hours by motor boat. From Teluk Batang you can reach Teluk Melano on a rented motor bike in 1 hour.
- Manara (Pontianak) ⟶ Rasau, takes 5 hours by motor boat. Then continue by car to Teluk Melano for 5 to 10 hours.

56. East Bornean Grey Gibbon

Hylobates funereus

Other common names: Uwa-uwa

Identification

The East bornean grey gibbon has a head and body length of around 48.5 cm for males and 47.5 to 49 cm for females. They weigh between 5 and 6.4 kg. The pelage of this species is dark brown or grey with a generally lighter color on the hands and feet.

Geographic Range

In Indonesia they are found in East Kalimantan Province and also elsewhere in north and northeastern Borneo.

Behavior and Ecology

As with other gibbons, this species is diurnal so is active during the day spending most of its time in trees. It is most active in the middle canopy, 25 to 30 m off the ground, usually for 8 to 10 hours each day. Most activity occurs in the morning and the evening. They generally rest in the afternoon. The diet of the East bornean grey gibbon is mainly fruit, especially figs. They are also known to consume leaves, flowers, and insects. There is little information available on the behavior and ecology of this species.

Conservation Status

This species is listed as Endangered on the IUCN Red list, and appears in CITES Appendix I.

Where to See It

The East bornean grey gibbon can be found from Sabah, Malaysia to the eastern part of Kalimantan, Indonesia. Sungai Wain Nature Reserve is located between Balikpapan and Samarinda. It takes only about 40 minutes by car to get there from Balikpapan and is one of the tourist attractions of Balikpapan. It covers 10,025 ha of primary and secondary forest. Other primates can also be found in this protected area including orangutans, proboscis monkeys, muller's gibbons, maroon langurs, long-tailed macaques, and lorises.

57. Abbott's Grey Gibbon

Hylobates abbotti

Other common names: Kelampiau, Penguak (Dayak)

Identification

Until recently, *H. abbotti* was thought to be a subspecies of *H. muelleri* but genetically and morphologically, it is now recognized as a distinct species (Ross et al., 2013). It hybridizes with *H. albabaris* in Central Borneo. There might be a morph or distinct subspecies in east-central Kalimantan although further research is needed to confirm or deny this. But currently, this species is assumed to be monotypic (Mittermeier et al., 2013). Information on specific body measurements are not available. But in general, weight is around 5.9 to 6.4 kg for males and 5.5 to 6 kg for females.

The body coloration of Abbott's Gray Gibbon is medium gray with a darker ventrum and crown. Males may be paler with a lighter eyebrow than females. The crown hair is directed fanwise from the front of the scalp and is markedly elongated over the ears. The feet and hands are not darkened.

Geographic Range

Distribution is limited in Indonesia to West Kalimantan near Sarawak.

Behavior and Ecology

They are active during the day and move arboreally by swinging using their arms through the tree canopy. This species can travel daily between 800 and 1,000 m with a home range average of 38 ha. Information on this species is still very limited and further research is needed but their diet undoubtedly includes young leaves, fruit, flowers, and insects.

Conservation Status

It is listed under CITES Appendix I and as Endangered on the IUCN Redlist. It is also protected under Indonesian and Malaysian law. Similar to other gibbons in Borneo, it is threatened by logging, hunting, and deforestation.

Where to See It

Abbott's Gray Gibbons can be seen in primary or secondary semi-deciduous monsoon, dipterocarp, and tropical evergreen forest. They can adapt to selective logging if sufficient tall, fruiting-bearing trees are retained. In West Kalimantan, they can be seen in several National Parks such as Danau Sentarum National Park, Betung Kerihun National Park, and several other conservation areas. In the two national parks mentioned above, you can also see primate species other than Abbot's grey gibbon.

- You can go to Danau Sentarum National Park. This park can be accessed by bus from Pontianak to Sintang, which takes 8 hours, or plane, which takes 1 hour. From Sintang to Semitau by minibus, takes 3 hours, by speedboat, 4 hours, or by long boat, 7 hours. From Semintau to Danau Sentarum by speedboat takes 1.5 hours. Alternatively, you can take a 2 hour plane ride from Pontianak to Puttusibau then continue to Nanga Suhaid by car or bus. Access to this park has become easier because the main road in West Kalimantan has been completed.
- The fastest access to this park is a 1.5 hour flight from Pontianak to Kapuas Hulu. Then you can rent a car and drive to the park in

4 to 5 hours. There are several small hotels available in a small city, Bada, close to the park.

- Betung Kerihun National Park. This national park is a little bit difficult to get to but can be reached through the following:
- From Pontianak to Sintang using a bus or plane. From Batu Layang bus station in Pontianak it will take between 7 and 8 hours. From Sintang, Puttusibau can be reach using a rental boat.

58. Siamang

Symphalangus syndactylus

Other common manes: Siamang, Kimbo (Indonesian

Identification

Adult Siamangs weigh approximately 11.2 kg. They are considerably larger than the gibbons in the hylobatid group, with an arm spread of as much as 1.5 meters. The head and body length is 80 to 90 cm, with no tail. The body and limbs are totally black except for the face, which is thinly clad with whitish hairs. The hair on the chin is denser and slightly tinged in brown. The throat sac is large, gray or pink in color and inflated during calls with crown fur laterally brushed that make it look like a flat crown. The second and third toes are fused.

Geographic Range

The Siamang is found in Indonesia (Barisan Mountains of west-central Sumatra), Malaysia (mountains of the Malay Peninsula south of the Perak River) and also in a small area of southern peninsular Thailand (Chivers,1974; O'Brien et al., 2003).

Behavior and Ecology

Siamangs are predominantly leaf eaters, eating less fruit than might be expected for their larger body size and relatively small territories. The proportional food preference is 50% leaves, 40% fruit, 5% flowers and 5% insects. This species is a spectacular brachiator swinging from tree to tree using only arms. They use all four limbs when climbing and are sometimes found walking bipedally on the ground and on branches.

A group of siamangs consists of one female, one male and several infants. Females are usually co-dominant with males and always lead the daily group travel. Males are more active socially than females. Intergroup interactions increase in frequency during periods of sexual activity, as offspring reach maturity, and during territorial behavior. Normally, sub adult and adolescent males participate in intergroup conflicts and accompany their fathers in chasing neighboring males (Chivers, 1974). As with *H. klossii*, *H agilis* and *H. lar* unmated males tend to travel alone looking for unmated females (Leighton, 1987).

Siamangs are normally active for about 10.5 hours a day. Commonly, their active time begins at dawn and continues until just before sunset. Usually, they use emergent trees to rest, sleep and sing. The main canopy is used for singing and foraging. They are rarely seen in the lower canopy. They travel about 800 to 900 m per day and will take around 6 days to cover their whole home ranges completely. Siamangs are highly territorial. Territories can reach up to 20 to 30 ha, which is smaller than the territories of other gibbons. As with other gibbons, they defend their territory through regular loud morning songs when encounters with neighbors or intruders occur.

Duet singing in this species can be heard almost daily. The duet usually begins mid morning or later (Chivers, 1974). Singing is less frequent on rainy or windy days and correlated with fruit abundance and mating activity (Chivers and Raemarkers, 1980). When singing, siamangs inflate the vocal sac which resonates with a deep boom. The song period begins with staccato announcement barks, in a phrase uttered by the male or by the pair in unison. During the warm-up the male contributes scream-chatters that eventually introduce the great call, sung by females five or six times. The great call opens with a long series of alternating booms and barks whose slight syncopation and steady acceleration impart excitement.

Conservation Status

The Siamang has lost around 66% of its original habitat, from 340,000 km² down to 120,000 km². Besides habitat loss, they are hunted for the illegal pet trade. Approximately 31,000 individuals remain in the wild. It is listed as endangered on the IUCN Redlist. To protect it, the Indonesian Government has some regulations and laws which are: the ministry of Agriculture decision letter 14 February 1973 no. 66/Kpts/um/2/1973, the ministry of forestry decision letter 10 June 1991 No. 301/Kpts-II/1991, and UU No. 5 year 1990.

Where to See It

They can be found in primary and secondary forest up to 3,800 m above sea level such as in Way Kambas National Park, Kerinci Seblat National Park, Bukit Barisan Selatan National Park, and Gunung Leuser National Park. The place to observe siamangs with the easiest access from Jakarta is Way Kambas National park which can be reached through:

• Flight from Soekarno-Hatta International Airport to Radin Inten II International Airport in Bandar Lampung. The flight takes around 30 minutes and is available every day.
  From Bandar Lampung you can go to Labuan Ratulama (Way Jepara) to Labuan Maninggai and then to Kuala Kambas using public transportation, which takes 1.5 hours.
  Alternatively, you can reach Way Kambas Nationa Park by rental car or public transportation to Teluk Betung, then to Sribawono and on to Way Jepara, which takes 2 to 3.5 hours. If you are using

a rental car, you can reach the National Park entry point through Plang Ijo. If you are using public transportation, from Way Jepara you can reach Plang Ijo by rental motor bike, which is a 15 minute ride.

- Bus and Ferry
  You can take a bus from Kalideres in West Jakarta to Merak Harbor, which takes around 2 to 3.5 hours. Take the ferry from Merak Harbour to Bakaheuni Harbor (2 hours). Then you will take a new bus or rent a small car to Rajabasa terminal bus station in Bandar Lampung. From the terminal you can take public transportation to Way Jepara , approximately 45 minutes, via the East Trans-Sumatra highway.

E. FAMILY POGIDAE (Gray, 1870)

The family Pongidae is considered to be the most advanced of the non-human primates. This family consists of 3 genera. Only the Orangutans of the genus Pongo are found in Indonesia. The two other genera, Gorilla and Pan, the Gorillas and Chimpanzees respectively, occupy the tropical or subtropical forests of Africa. The species of Gorilla and Pan have different lifestyles compared to Indonesia's Pongo. They both live in groups with Alpha males while Orangutans are solitary.

Pongo: Orangutans

This great ape is only found on the island of Sumatera and Borneo. Historically the Sumatran and Bornean orangutans were long considered to be subspecies of Pongo pygmaeus, but in the taxonomic reviews by Groves (2001) and Brandon-Jones *et al.*, (2004) it was argued that the evidence supports the acceptance of the Sumatran orangutan, *Pongo abelii,* as a distinct species separate from its Bornean relative, *Pongo pygmaeus.* This new classification has since been widely adopted and was recognized at the most recent orangutan Population Habitat Viability Analysis (PHVA) workshop (Singleton *et al.,* 2004). In addition, the Bornean species is divided into three subspecies, *P. p. pygmaeus, P. p. mario* and *P. p. wurmbii.* Nater et al., (2017) published the description of this new species, based on phylogenetic and morphological analyses very recently. Indications of the uniqueness of the Tapanuli population came from the skeletal material of an adult male orangutan killed in 2013. When compared to other skulls it turned out that certain characteristics of the teeth and skull of the

© The Author(s), under exclusive license to Springer Nature Switzerland AG 2022
J. Supriatna, *Field Guide to the Primates of Indonesia,*
https://doi.org/10.1007/978-3-030-83206-3_8

Tapanuli orangutan were different. Anton Nurcahyo, an Indonesia graduate student who studied the specimens of orangutan from the Tapanuli area found that the skull and jaws were softer than those of the Sumatran and Bornean Orangutans. This Tapanuli orangutan has been named as *Pongo tapanuliensis* (Nater et al., 2017).

Orangutans are more arboreal and more solitary than the other apes. Though the males travel on the ground more than the females, they all live in trees and use their grasping hands and feet to climb slowly and to suspend themselves from tree branches to feed. Orangutans feed on some ripe fruit and, after a day of gracefully living in fine balance with their tropical forest home, they make their nest and go to bed. Orangutans are the largest primarily arboreal mammals, with males averaging 80 kg and females 40 kg (MacKinnon, 1974; Markham and Groves, 1990), orangutans have few nonhuman predators; tigers are a potential terrestrial predator in Sumatra, but on Borneo there are currently no cats, snakes, or raptors large enough to pose a serious threat (Hart, 2007). Orangutans prefer a diet of ripe fruit, although when necessary they exploit a wide range of low-quality including leaves, pith, and inner bark (MacKinnon, 1974; Knott, 1998). With few predators, the density of orangutans is limited by the relative scarcity of their key food sources. As a result, populations are typically at or near carrying capacity (Marshall *et al.*, 2009), with densities of only 1–3 animals/km^2 (Husson *et al.*, 2009).

This species has long been protected by law in Indonesia, but it is still sometimes taken and exported illegally. The most recent estimates of Sumatran orangutan numbers are around 7,300 individuals (Singleton et al., 2004) occupying forests that cover a total of 20,552 km², but only those regions below 1,000 m above sea level (about 8,992 km²) are considered to harbor permanent populations.

In addition to the truly wild populations, a new population is being established in the Bukit Tigapuluh National Park in Jambi and Riau Provinces via the re-introduction of confiscated illegal pets. This population currently numbers around 70 individuals and is already reproducing.

59. Sumatran Orangutan

Pongo abelii

Other common names: Mawas (Indonesian)

Identification

Sumatran Orangutans display extreem sexual dimorphism. Some adult males develop cheek pads, called flanges, beginning at 10 years of age, whereas others may remain unflanged for up to 20 years. These cheeck pads are flat and covered with downy hair, tend to be more muscular than other orangutan species. Adult males also have a throat sac for producing loud calls to announce their presence (Galdikas, 1985b; MacKinnon, 1974; Rijksen, 1978; Rodman, 1984). The adult males of *P. abelii* have a longer beard and mustache compared with females, which also have a flatter face that is more elongated and zero-shaped.

The male's body size is twice that of females. Male head and body length is between 94 and 99 cm while female head and body length is between 68 and 85 cm. The average body weight for flanged males is between 60 and 85 kg, while males without flanged cheeks weigh between 30 and 65 kg, and females between 30 and 45 kg. Males can exceed 100 kg in weight when they get older (Mittermeier et al., 2013). Sumatran Orangutans are slimmer with a longer face compared to Bornean Orangutans. In contrast to the Bornean species, Sumatran Orangutans have fairer, longer, denser and more fleecy hair with black-haired orangutans seemingly rarer than in Borneo; when it does occur, the black hair often has a reddish tinge.

Geographic Range

This species is endemic to the island of Sumatra, Indonesia. It is generally restricted to the northern part of the island (Wich et al., 2003). There is evidence that it was once far more widespread and that populations still occurred as far south as Jambi until at least the mid 1800's (see Rijksen, 1978) and there have even been reports of its existence in some parts of West Sumatra province as recently as the 1960's.

Today, the majority of wild Sumatran Orangutans can be found in the province of Aceh (more formally known as Nanggroe Aceh Darussalam, or NAD), located at the northernmost tip of the island. There are orangutan populations within North Sumatra Province, but the largest of these also straddle the border with Aceh. Within Aceh, almost all remaining forest patches of any size still support populations of orangutans at the lower altitudes. However, many of the mountains exceed 1,000 m in altitude, and so large tracts of forest contain few or no orangutans.

Behavior and Ecology

The Sumatran Orangutan is almost exclusively arboreal except for occasional forays on the ground. Females virtually never travel on the ground. Large males have been observed using the ground for travel, most likely because the trees could not support their weight. This is in contrast to Borneo, where orangutans (especially adult males) will more often descend to the ground. Orangutans on both islands depend on high quality primary forests. Whilst Bornean orangutans appear to be able to tolerate habitat disturbance better, in Sumatra they tend to plummet by as much as 60% with even careful selective logging (Rao and van Schaik, 1997).

Sumatran Orangutans are primarily frugivores, but they also eat significant amounts of leaves, and insects (termites and ants). On occasions, they have been observed to consume meat such as slow lorises (Utami & van Hoof, 1997; Fox et al., 2004, Wich et al., 2006). Females do not give birth until around 15 years of age (Wich et al., 2004). Interbirth intervals among Sumatran orangutans, at 8.2 to 9.3 years, are longer than in Borneo (6.1to 7.7 years; Wich et al., 2004; van Noordwijk and van Schaik 2005) and gestation lasts approximately 254 days (Kingsley, 1981). Longevity in the wild has been estimated at 58 years for males and 53 years for females (Wich et al., 2004).

Orangutan males exhibit bi-maturism, this means some orangutans have grown fully developed flanged and others have smaller or unflanged. Both flabfed and unflanged males are capable of reproducing, but employ differing mating strategies to do so (Utami et al., 2002). Flang development is assumed to be triggered by the level of stress. Where there are flanged males in an area, while flang development in other males is inhibited or unflanged. Home ranges for females have been shown to be at least 800 ha and possibly up to 1,500 ha in some areas. The true extent of home range sizes for males is still not known, although ranges in excess of 3,000 ha can easily be inferred (Singleton and van Schaik, 2001).

Conservation Status

This species is listed on Appendix I of CITES and as Critically Endangered on the IUCN Redlist. It is strictly protected under Indonesian domestic legislation (UU No 5/1990). The protection in conservation areas of large tracts of primary forest at altitudes below 1,000 m above sea level is needed to secure a long term future for this species. A major stronghold of the Sumatran orangutan is within the forests of the Leuser Ecosystem conservation area. This area of 2.6 million hectares supports about 75% of remaining Sumatran orangutans. The Leuser Ecosystem was inaugurated by Presidential Decree in 1998 and its conservation is called for in the Act of Parliament No 11/2006 concerning Governance in Aceh. Management of the Ecosystem does not exclude non-forest uses, but stresses the importance of managing the area in a sustainable fashion with conservation of natural resources as the primary goal. Within the Leuser Ecosystem is the designated (approximately 900,000 hectares). The Gunung Leuser National Park, which is mostly high mountains and therefore supports only around 25% of the remaining orangutans on the island. It is also a Man and Biosphere reserve and part of the Tropical Rainforest Heritage of the Sumatra World Heritage Cluster Site. Outside of this park there are no other notable large conservation areas harboring this species.

As of the end of 2006, efforts are also underway, in the wake of the December 2004 tsunami, to establish a second protected area in the Ulu Masen ecosystem in the north-east and north-west of Aceh Province, along similar lines to the Leuser Ecosystem. However, this process is still in its early stages and there are already threats to open up at least four large logging concessions in this area. Overall t has been an estimated decline of over 80% over the last 75 years (assuming a generation length of at least 25 years). The decline of the species continues, as the forests within its range are under major threat. This species is seriously threatened by logging (both legal and illegal), wholesale conversion of forests to agricultural land and oil palm plantations, and fragmentation by roads. Animals are also illegally hunted and captured for the international pet trade but this appears to be more a symptom of the above, and includes orangutans being killed as pests as they raid fruit crops at the forest edge. Current major issues threatening Sumatran orangutans include the road network in Aceh province, which if legitimized by the government will rapidly fragment most of the remaining population. Another major concern is the re-issuing of logging permits for large tracts of forest in Aceh. During the late 1990's an assessment of forest loss carried out by van Schaik *et al.,* (2001) concluded that forests supporting at least 1,000 orangutans were being lost each year within the Leuser Ecosystem alone. Fortunately for the orangutans, these loss rates subsequently reduced dramatically as a result of a major civil conflict in the province, and the imposition of a moratorium on logging in Aceh. In 2005, however, a peace deal was negotiated between the Indonesian Government and the Acehnese separatist rebels. This has led to a new period of political stability in the province and many new applications to open up logging concessions and plant palm oil estates in orangutan habitat areas. In parts of North Sumatra orangutans are also still hunted on occasions for food. Most orangutans are found outside of protected areas, including within potential logging areas and conversion forests. After a period of relative stability, pressure is increasing once again on these forests as a direct result of the recent peace accord and due to a dramatic increase in demand for timber and other resources after the December 2004 tsunami.

Where to See It

It is not difficult to find this primate. There are places that are easy to reach for this purpose, where both wild and semi-wild individuals can be seen. They are easy to find in Bohorok or Kutacane, both places are inside the Gunung Leuser National Park, where there are facilities for re-introducing ex-pet orangutans to the wild. There is also an orangutan research center.

60. Bornean Orangutan

Pongo pygmaeus

Other common names: Orangutan, Orang hutan (Indonesian)

Identification

Weight range is between 60 and 85 kg for flanged males, from 30 to 65 kg for unflanged males, and from 30 to 45 kg for females. Adult head and body length is between 72 and 97 cm. The ears are small, without lobes and the nose is small. They have a high forehead with a bulging snout. The mouth has thin lips. Legs are short and relatively weak, but they have very powerful hands and arms. The arms can reach to the ankles when this animal is erect.

As for Sumatran Orangutans, Bornean Orangutans display extreem sexual dimorphism. Adult males can reach up to 100 kg. They also exhibit bimaturism, which means that there are two adult male morphologies, one with a flang and one unflanged (Utami et al., 2002). A highly developed throat sac in flanged individuals also generates a noticeable double chin. Bornean Orangutans have dark hair. Generally, their body coloration is dark rufous or reddish brown and is rather thin and shaggy. They also have a prominent pale skinned eye-ring throughout adolenscence. In infants, the body coloration is usually pinkish but will darken later on.

Geographic Range

Roos et al., (2014) and Singleton (2004) recognized 3 subspecies of Bornean Orangutan:

- *P. p. pygmaeus*

This subspecies is found in the north of Borneo Island. In Indonesia, it is found in northwetern Kalimantan, but it also occurs in Sarawak

- *P. p. morio*

This subspecies occurs in southern and Central Kalimantan.

- *P. p. wurmbii*

P. p. wurmbii is the northeast Bornean Orangutan which is found in East and North Kalimantan and Sabah.

Behavior and Ecology

Bornean Orangutans are frugivorous with a high proportion of wild figs in their diet. Many other kinds of vegetation, as well as insects, probably small vertebrates and eggs are also consumed. Food is plucked with one hand, usually between fingers and palm because the thumb is too short for efficient manipulation. They are usually silent chewing fruit, leaves and other vegetable matter. At Tanjung Puting National Park, Orangutans eat over 200 varieties of wild fruits. In total, they are known to eat over 400 types of food.

Unlike most other higher primates, adult orangutans are semi solitary or solitary with adult females accompanied only by their depended young. The young will then learn about which foods are safe to eat from observing their mother and sharing food with her. They will also learn to build a mental map of the forest habitat of their mother's home range. Infant care will last from 7 to 8 years, after which both females and males will disperse to a solitary existence except young and mother with infants..

Normally Bornean Orangutans move by climbing through large trees and swinging from branch to branch. Jumping is not seen in this species. Movements commonly are unhurried. They do not have the ability to swim. On the ground they walk quadrupedally or bipedally but it is rarely seen. Their daily movements is from 200 to 1,000 m.

Two or more females, usually with infants, are associated in stable social groups. Their home ranges usually overlap with 3 – 4 other individuals in the same area. They don't have permanent territories but adult males occupy larger home ranges than adult females. Adult males have home ranges of between 2 to 6 km² that overlap with the ranges of several adult females. Whilst adult males tend to travel alone, sub-adults have been seen to travel together. Males are not aggressive to one another. Activity peaks in the morning and late afternoon with a rest period around the middle of the day. Orangutans are comparatively quiet. They communicate with sounds made by smacking the lips. In anger or frustration, orangutans grind their teeth. Night-sleeping occurs in nests built in trees, usually 12 to 19 m above ground. Nest shapes are simple, with little branch weaving and are seldom re-used. Orangutans lie on the side or back to sleep. Population densities range from 0.2 to 5.0 individuals/km².

For mating, males usually prefer fully adult females and females prefer dominant adult males. Copulation commonly occurs with the partner's front to front, hanging by the arms from a support and it is not necessarily confined to estrus. Predators are unknown, except for man.

Conservation Status

Because orangutans consume food that mostly comes from trees, they are very sensitive to changes in their forest habitat. Bornean Orangutans have lost more than half of their original habitat from 415,000 km² to only about 165,000 km² today. They are also found outside protected areas, where they are more vulnerable, in forest that has already been exploited for timber, coal, and gold or been converted to oil palm plantations.

Threats also come from hunting for meat and from the pet trade. Only the young are popular as pets so mothers are killed simply for one offspring. This has hastened the decline of the orangutan population in the wild. As they are also found near plantations, where they are considered to be pests, Bornean Orangutans are often shot or tortured by crop owners. In some areas these practices have already caused local extinction of this species. To maintain the existence of orangutans in the wild, this great ape has been protected since the 1931 wildlife protection regulation 1931 No. 233 and the decision letter of Ministry of Forestry 10 June 1991 No. 301/Kpts-II/1991 and UU no 5 1990. Bornean Orangutans are protected by CITES Appendix I and in 2016, the IUCN classified Bornean orangutans as critically endangered due to a precipitous population decline caused by the destruction of their forest habitat for palm oil and pulpwood plantations.

Where to See It

Primary or secondary forest, from lowland swamps to upland forest as high as 1500 m above sea level. They are now also found as isolated populations in small fragmented forest remnants. While mountains and large rivers have been barriers to their distribution, today the main barrier is forest destruction (Mittermeier *et al.,* 2013).

The best place to see Bornean orangutans is in Tanjung Puting National Park in Central Kalimantan as well as many others such as Kutai National Park in East Kalimantan and also Gunung Palung National Park in West Kalimantan. To get to Tanjung Puting, you can fly from Jakarta to Pangkalan Bun city in Central Kalimantan, a flight of approximately 1 hr. From this airport you can rent a car to Kumai where you can rent a boat to the park. Tanjung Puting National Park is the best place to see orangutans in the wild because for more than 40 years, Dr. Birute Galdikas has been studying the wild populations in the park. There are also many hotels, motels, and homestays in the small town of Kumai or Pangkalan Bun, both close to the park. The Rimba lodge hotel, which is a very beautiful place when you wake up in the morning to see rivers and jungle at the same time. Another interesting accommodation option is a boat that will provide a bed and dinner as you travel to the park.

61. Tapanuli orangutan

Pongo tapanuliensis

Other common name: Mawas Tapanuli (North Sumatra)

Identification:

Nater et al., (2017) notice differences in other morphological features. The hair is frizzy and very thick or dense compared to Sumatran and Bornean orangutans. Males have long whiskers and hairy beards. Cheek pads are flat with a lot of pale hair. Females also grow beards. Mathematical modelling and DNA analyses suggested that Tapanuli orangutans split from their Sumatran and Bornean cousins about 3.4 million years ago. The great call of the Tapanuli orangutan has a higher frequency and lasts longer than the great calls of the other two species.

Geographic range

Tapanuli orangutans are confined to an area of about 1,100 square kilometers (425 square miles) in the Batang Toru forest in the South, central and North of Tapanuli districts of Northern Sumatra

Province. There are thought to be fewer than 800 individuals, which makes this species the rarest great ape.

Behavior and Ecology

Orangutan density varies with elevation and habitat availability (Delgado and van Schaik, 2000). According to Buij *et al.,* (2002), even though there is forest clearance in Sumatra, there are still some areas assumed to be suitable for orangutan habitat in the future. Forests of Batang Toru, especially the lowland and mixed forests, have been highly fragmented due to logging, road opening, and land-use change for agriculture. Because of this, the density of orangutans in Batang Toru Watershed tends to increase in line with the increase in elevation. With decreasing number of fruit trees and other sources of food in lowland forest, the orangutan will move to higher elevations in search of suitable habitat and sources of food.

A population survey of the orang-utans at Batang Toru watershed in North Sumatera, Indonesia was conducted over a period of 10 months from November 2005 to September 2006 (Sitaprasasti, 2007). Through three extensive and broad surveys in 16 locations with 40.6 km transect length (varies between 750-1,500 meters above sea level), also regular monthly monitoring in existing transects in five model sites (Lobu Pining, Sibulan-bulan, Sipetang, Sitandiang and Uluala) throughout three regencies, it was estimated that the orangutan density was between 0.2-0.82 individual/km² distributed in a landscape unit of 74.886 ha, which is known as a potential orangutan habitat over the total area of 90,000 ha. The nest count and line transect methods were used in determining population density, and the verification of orangutan presence was done through nest sightings, direct sightings and vocalizations. The survey result showed population densities were found to be higher in the old, moist secondary forest (0.82 individuals/km²) compared to those in the mixed forest (0.26 individuals/km^2). Habitat disturbance caused by forest conversion is believed to reduce the orangutan density within mixed and lowland forests. Habitat disturbance caused by land conversion seemed to affect the orangutan density within mixed and lowland forests Orangutan density in West Batang Toru forest is lower in comparison with the other forest area in North Sumatera, for instance in Aceh with density >6 individuals/km^2 (van Schaik *et al.*, 1995). The orangutan distribution pattern is highly affected by food resources availability, altitude, river shed and human economic activities in the habitat of orangutans.

Conservation Status

Tapanuli orangutan has not been categorized by IUCN but since the population is only 800 individuals left living in the upland forest regions of the island. Since Orangutan Sumatra has been categorized as critically endangered species, this species of course is automatecally placed as critically endangered species. This species is already at a high risk of extinction. It is facing further pressure from the construction of a hydropower plant, earmarked for completion in 2022, which could flood up to 8 percent of the Tapanuli orangutan's habitat. It could also cut off forest corridors used by orangutans to move between populations, leading to more isolation and inbreeding. The Batang Toru Forest consists of two forest blocks called West and East Batang Toru or Sarulla Forest Range. The Batang Toru Forest is located in North Sumatra Province south of the second world largest lake of Lake Toba. Roads separate West Batang from the East Sarulla area, in which orangutans also are found. Geographically, the Batang Toru Forest is located at 98^0 50' - 99^{0}18' East Longitude and 1^{0}26' - 1^{0}56' North Latitude. This landscape is predicted to have suitable habitat for orangutans in the southern part population of North Sumatra

Province (Rijksen and Meijaard, 1999; Wich et al., 2003; Djojoasmoro et al., 2004).

Orangutan habitat in Batang Toru Forest is also important water catchments area that encompasses four regencies: North Tapanuli, Central Tapanuli, Sibolga and South Tapanuli. Primary rainforest dominates the vegetation cover, which grows on steep hillsides with more than a 60-degree slope and mountainous area with the highest peat at Mt. Lubuk Raya (1.856 meters above sea level). The region is mountainous and the results of historic volcanic activity called Toba Super Volcano and the formation of geology is Volcanic Toba Tuff as the dominant geology rock type. Soil type is dominated by Podsolic Red-Yellow and Aluvial (Perbatakusuma, et al, 2008)

The Batang Toru Forest Block holds at least six principal habitat types including moss forest (above 600 meters), hillside moist forest (dominant between 200 m-600 m), lowland, cliffs and talus slopes, secondary forest, and riparian forest. Forest landscape of Batang Toru has 234,399 ha which included four districts Tapanuli Utara (North Tapanuli), Sibolga, Tapanuli Tengah (Middle Tapanuli), and Tapanuli Selatan (South Tapanuli). Based on satellite imagery classification that was analyzed by Conservation International Indonesia, primary forest cover in 2000 was 148.000 ha with infrastructure undeveloped in frontier areas. The area includes of mix of official forest status and land use classifications such as nature reserve, protection forest and production forest. The forest area is covered by five big watershed: Batang Toru with 92.121 ha wide, Aek Kolang 42.663 ha, Watershed Bila, Barumun, and Batang Gadis (Perbatakusuma, et al., 2008)

International in 2006 and other institution revealed that the Batang Toru and adjacent areas are home to a rich variety of the Sumatran species, particularly mammals, birds and plants, which are globally threatened. Sixty-seven species of mammals, two hundred eighty-seven birds and one hundred ten herpetofauna have been recorded in the area. Of this total number of mammals species, twenty species are protected under Indonesian law and twelve are globally threatened. Among these are Sumatran Orangutan (*Pongo abelli*), Sumatran tiger (*Panthera tigris sumatrae*), serow (Capricornis sumatrensis), Malayan tapir (*Tapirus indicus*), Malayan sun bear (*Helarctos malayanus*), slow loris (*Nycticebus coucang*), Golden Cat (*Pardofelis marmomata*). The survey also discovered rich avifauna diversity in the region, including rare as well as threatened species. Of this total number of bird species fifty-one species are protected under Indonesian law and sixty-one are globally threatened, such as Sunda Blue Flycatcher (*Cyornis caerulatus*), Wallace's Hawk-eagle (*Spizaetus nanus*), Blackcrowned (*Pitta venusta*). Initial data from the Batang Toru suggest that it holds some of the highest levels of vascular plant biodiversity, with 688 different species. Of this total number of plant species, 138 species of orangutan food resources, 8 species globally threatened, including Nepenthes sumatrana (Miq.), the largest flower in the world *Rafflesia gadutensis* Meijer, and the tallest flower in the world *Amorphophalus baccari* and *Amorphophalus gigas* (Perbatakusuma, et al., 2008).

Immediate threat for Tapanuli orangutan is hunting. The habitat of this orangutan is easy to get access it means also it will be more vulnerable than other orangutan species. Any loss of this orangutan will significantly change the orangutan survival.

Where to see it

Tapanuli orangutans can be seen in the forests of Batang Toru and in some agroforestry gardens belonging to local communities in the region. They are sometimes seen eating durian fruits. Batang toru can be reached by plane from Jakarta to Sibolga or from Jakarta to Silangit airport. From there you can rent a car to drive to Batang toru. Hotels or resorts are available in Sibolga or in Padang Sidempuan.

There are many roads that can take you to Batang Toru and you can ask local people if there are orangutans in their gardens. There is also a community tourism organization in Aek Naboru in South Tapanuli Regency, which has developed an ecotourism package. You can also contact SOCP (Sumatran Orangutan Conservation Program) in Medan (search for the website).

Figure 10. Three species of Indonesian orang-utan: a. Bornean Orangutan left: *Pongo pygmaeus wurmbii* (Kristana Makur), center: *Pongo pygmaeus wurmbii* (FX Ngindang), right: *Pongo pygmaeus morio* (Misdi); b. Sumatran Orangutan *Pongo abelii* (Misdi & Sri Suci Utami Atmoko); c. *Pongo tapanuliensis* (Kuswanda);

References

Abegg, C. and B. Thierry. 2002. The phylogeny status of Siberut macaques: hints from the bare teeth display. *Primate Report*, 63:73-78.

Aimi, H. and A. Bakar. 1992. Taxonomy and distribution of Presbytis melalophos group in Sumatra, Indonesia. *Primates,* 33:191-206.

Aimi, H. and A. Bakar 1996. Distribution and development of Presbytis melalophos group in Sumatra, Indonesia. *Primates* 37:399-499.

Ampeng, A. & B.M. Md. Zain. 2012. Ranging patterns of critically endangered colobine, Presbytis chrysomelas chrysomelas. *The Scientific World Journal* 2012:7.

Andayani, N.W. Y. Brockleman, T. Geissmann, V. Nijman, J. Supriatna. 2008. Hylobates moloch. In: IUCN Redlist of threatened species, Version 2011. 2. url:http/www. iucn redlist. org

Andayani, N. , J. C. Morales, M. R. J. Forstner, J. Supriatna, and D. J. Melnick. . 2001. Genetic Variability in mt DNA of the Silvery Gibbon: Implications for the Conservation of a Critically Endangered Species. *Conservation Biology,* 15 (3):770-775.

Ankel-Simons, E. 2000. *Primate Anatomy: An Introduction. 2nd edition. Academic Press, California.*

Ario, A. , A. P. Kartono, L. B. Prasetyo, and J. Supriatna. 2018. Habitat Suitability of Release Site for Javan Gibbon (Hylobates moloch) In Mount Malabar Protected Forests, West Java. *Jurnal Manajemen Hutan Tropika* 24, (2):92 101,

Arisona, J. 2008. Studi Populasi, Perilaku, dan Ekologi kukang Jawa (Nycticebus javanicus E. Geoffrey 1812) di hutan Badogol Taman Nasional Gunung Gede Pangrango, Jawa Barat. Tesis Sarjana Biologi, Universitas Indonesia.

Azuma, S. , A. Suzuki, R. Ruhiyat. 1984. The distribution of primates in Sebulu and R. Mahakam. Kyoto Univ Overseas *Res Rep Stud Asian Non-Human primates,* 3:45-54.

Bartlett, T. Q. 2009. The gibbons of Khao Yai: seasonal variation in behavior and ecology. Upper Saddle River. Pearson Prentice Hall, p. 170.

Bennett, E. L. and Davies, A. G. 1994. The ecology of colobines, pp. 129-172. In: Davis and Oates.

Bennett, E. L. and Sebastian, A. C. 1988. Social organization of proboscis monkey (Nasalis larvatus) in mixed coastal forest in Sarawak. Int. *J. Primatol.,* 9:233-255.

Bersacola, E. , D. A. , Ehlers Smith, W. J. Sastramidjaja, Y. Rayadin, and S. M. Cheyne. 2014. Population density of Presbytis rubicunda in small primary dipterocarp forest in east Kalimantan. Asian Primates Journal 4 (2):16-26.

Bernstein, L. S. 1968. The lutong of Kuala Selangor. *Behaviour* 32:1-16

Boonratana, R. 2000. Ranging behaviouJr of proboscis monkey (Nasalis larvatus) in the lower Kinabatangan, Northern Borneo. *Int. J. Primatol.,* 21:497-518.

Brandon-Jones, D. 1995. Presbytis fredericae (Sody 1930), an endangered colobine species endemic to central Java, Indonesia. Primate Conserv 16:68-70.

Brandon-Jones, D. 2004. A taxonomic revision of the langurs and leaf monkeys (Primates: Colobinae) of South Asia. *Zoos' Print Journal,* 19(8):1552-1594.

Brandon-Jones, D. , A. A. Eudey, T. Geissmann, C. P. Groves, D. J. Melnick, J. C. Morales, M. Shekelle, and C. B. Stewart. 2004. Asian primate classification. *International Journal of Primatology,* 25(1):97- 164.

Baker, S. , A. Kohlhaas, J. Supriatna, J. Sugardjito, C. Southwick, B. E. Mulligan & E. Erwin. 1989. Field observations of the behavior of Macaca nigrescens. *American Journal of Primatology,* 18(2):133.

Bismark, M. 2010. Proboscis monkey (Nasalis larvatus): Bioecology and Conservation, pp. 217-233". In: Indonesia Primates; Developments in Primatology: Progress and Prospects (Gursky-Doyen & Supriatna). Springer, Chicago.

Brandon-Jones, D. , A. A. Eudey, T. Geissmann, C. P. Groves, D. J. Melnick, J. C. Morales, M. Shekelle and C. -B. Stewart, 2004. Asian primate classification. *Int. J. Primatol.*, 25:97–164.

Brockelman, W. Y. 1985. A gibbon pelt (Hylobates lar entelliodes) from Khao Yai National Park, Saraburi Province, Thailand. *Nat His Bull Siam Soc.*, 33:55-57.

Brockelman, W. Y 2004. Inheritance and selective effects color phase of White-handed gibbons (Hylobates lar) in central Thailand. *Mammal Biol,* 69:73-80.

Buckley, C. 2004. Survey of Hylobates agilis albibarbis in unprotected peat swamp forest: Sebangau Catchment Area, central Kalimantan. M. Sc. thesis, Oxford Brookes University, Oxford.

Buckley, C. , K. A. I. Nekaris, and S. Husson. 2006. Survey of Hylobates agilis albibarbis in logged peat swamped forest: Sebangau catchment, central Kalimantan. *Primates,* 47:327-335.

Chaeril, A. H. Suci, M. Nurhidayat, R. W. Langodai, and Sulkarnaen. 2011. Identifikasi dan Pemetaan Sebaran Tarsius di SPTN Wilayah I Balocci Taman Nasional Bantimurung Bulusaraung. Bantimurung : Balai Taman Nasional Bantimurung Bulusaraung.

Chen, J, D. Pan, C. P. Grooves, Y. Wang, E. Narushima, H. Fitch-Sneider, P. Crow, N. T. Vu, O. Ryder, H. Zhang, Y. Fu, & Y. Zhang. 2003. Molecular phylogeny of Nycticebus inferred from mitochondrial genes. *Int. J. Primatol.*, 27:1187-1200.

Chivers, D. J. 1974. The siamang in Malaya: A field study of a primate in tropical rain forest. Contrib. *Primatol,* 4:1-355. Basel: S. Karger.

Chivers, D. J. 1986. Southeast Asian primates in: Bernischke (1986).

Chivers, D. J. and K. M. Burton 1988. Some observation on the primates of Kalimantan tengah, Indonesia. *Primate Conservation,* 9:138-146.

Chivers, D. J. and J. J. Raemarkers1980. *Longterm change in behavior,* pp. 209-260 in: Chivers

Corbet, G. B. & J. E. Hill. 1992. *Mammals of the Indomalayan region: A systematic Review.* Oxford University Press, Oxford.

Crockett, C. M. and W. L. Wilson. 1980. The ecological separation of Macaca nemestrina and Macaca fascicularis in Sumatra. in Lindburg.

Curtin, S. H. 1980. Dusky and banded leaf monkeys. In DJ Chivers (ed.): *Malayan Forest Primates.* New York: Plenum Press, pp. 107-145.

Davies, A. G. 1984. An Ecological studies of *Presbytis rubicunda* in the Dipterocarp Forests of Sabah, Northern Borneo. PhD dissertation, University of Cambridge, UK.

Davies, A. G. 1987. Adult male replacement and group formation of Presbytis rubicunda. *Folia Primatol,* 49:111-114.

Davies, A. G. 1991. Seed-eating by red leaf monkeys (Presbytis rubicunda) in dipterocarp forest of northern Borneo. *Int. J. Primatol.,* 12:119-144.

Davies, A. G. 1994. Colobine populations. In AG Davies and JF Oates (eds.): Colobine Monkeys: Their Ecology, Behaviour and Evolution. Cambridge: Cambridge University Press, pp. 285-310.

Davies, A. G. , E. L. Bennett, and P. G. Waterman. 1988. Food selection by two South-east Asian colobine monkeys (Presbytis rubicunda and Presbytis melalophos) in relation to plant chemistry. *Biol. J. Linnean Soc.,* 34:33-56.

Davies, A. G. & Baillie 1998. Soil eating by red leaf monkey (Presbytis rubicunda) in Sabah, Northern Borneo. *Biotropica,* 20:252-258

Driller, C. , S. Merker, D. Perwitasari-Farajallah, W. Sinaga, N. Anggraeni and H. Zischler. 2015. Stop and go–waves of tarsier dispersal mirror the genesis of Sulawesi Island. PLoS ONE 10(11) e0141212. DOI: 10. 1371/journal. pone. 0141212

Ehlers Smith, D. A. , Y. C. Ehlers Smith, and S. M. Cheyne. 2013a. Home-range use and activity patterns of the Red Langur (Presbytis rubicunda) in Sabangau Tropical Peat-Swamp Forest, Central Kalimantan, Indonesian Borneo. *International Journal of Primatology,* 34:957-972.

Ehlers Smith, D. A. , S. J. Husson, Y. C. Ehlers Smith, and M. E. Harrison. 2013b. Feeding ecology of red langurs in Sabangau Tropical Peat-Swamp Forest, Indonesian Borneo: frugivory in a non-masting forest. *American Journal of Primatology,* 75:848-859.

Eudey, A. A. 1987. IUCN/SSC Primate Specialist Group Action Plan for Asian Primate Conservation. 1987-1991. IUCN/SSC Primate Specialist Group and World Wildlife Fund US, Washington, DC.

Fleagle, J. G. 1999. *Primate adaptation and evolution.* 2nd edition, Academic Press, San Diego, California.

Fooden, J. 1969. *Taxonomic and Evolution of the monkey of Celebes* (Primates, Cercopithecidae). S. Karger, Basel.

Fox, E. A. , C. van Schaik, A. Sitompul, & D. N. Wright. 2004. Intra-and Inter-populational differences in orangutan (Pongo pygmaeus) activity and diet: implication for invention of tool use. *Am. J. Physic. Anthrop,* 123:162-174.

Froehlich, J. and J. Supriatna. 1996. Secondary introgression of M. Maurus and M. tonkeana at South Sulawesi and the species status of *Macaca togeanus.* In J. E. Fa, and D. G. Lindburg. *Evolution and Ecology of Macaques Societies,* Cambridge University Press, pp. 43-70.

Fuentes, A and Tenaza, R. R 1995. Infant parking in the pig tailed langur (*Simias concolor*). *Folia Primatol,* 65:172-173.

Fuentes, A. 1996. Feeding and ranging in the Mentawai island langur (*Presbytis potenziani*). *Int. J. Primatol.,* 17:525-548.

Fuentes, A. . 1977. Feeding and ranging in the Mentawai island langur (*Presbytis potenziani*). *Int. J. Primatol.,* 17:525–548. doi: 10. 1007/ BF02735190.

Fuentes, A. . 1977. Current status and future viability for the Mentawai Primates. *Primate Conserv.,* 17:111–116.

Fuentes, A. and M. Olson. 1995. Preliminary observations and status of the Pagai macaque. *Asian Primates,* 4:1-4.

Furuya, Y. 1961-2. The social life of silvered leaf monkeys. *Primates,* 3(2), 41-60

Galdikas, B. M. F. 1985 Adult Male Sociality and Reproductive Tactics among Orangutans at Tanjung Putting. *Folia Primatol,* 45:9–24. https://doi. org/10. 1159/000156188

Galdikas, B. M. F. 1985b. Subadult male orangutan sociality and reproductive behavior in Tanjung Putting. *Am J. Primatol.,* 8:87-99

Galdikas, B. M. F. 1988. Orangutan diet, range, and activity at Tanjung Putting, Central Borneo. *Int. J. Primatol.,* 9:1-35.

Geissmann, T. 2002. Taxonomy and evolution of gibbons. *Evol. Anthropol.,* 11:28–31.

Geissmann T. 1995. Gibbon systematics and species identification. *International Zoo News,* 42:467-501.

Geissmann, T. , S. Bohlen-Eyring, and A. Heuck. 2005. The male song of the Javan silvery gibbon (*Hylobates moloch*). *Contributions to Zoology,* 74 (1/2):1-25.

Geissman, T. and V. Nijman. 2006. Calling in Wild Silvery Gibbons (*Hylobates moloch*) in Java (Indonesia): Behavior, Phylogeny, and Conservation. *American Journal of Primatology,* 68:1-19. Wiley Liss, Inc.

Geissmann, T. & Nijman, V. 2008. Hylobates agilis. The IUCN Red List of Threatened Species 2008: e. T10543A3198943. http://dx. doi. org/10. 2305/IUCN. UK. 2008. RLTS. T10543A3198943. en. Downloaded on 22 February 2017.

Goodman, S. M. 1989. Predation by Grey leaf monkey (*Presbytis hosei*) on the content of a Bird's nest at Mt Kinabalu, Sabah. *Primates,* 30(1):27-29.

Gittins, S. P. 1980. Territorial behavior in the agile gibbons. *Int. J. Primatol.,* 1:381-399

Gittins, S. P. 1982. Feeding and ranging on the agile gibbon, *Folia Primatol.,* 38:39-71.

Gittins, S. P. and J. J. Raemarkers. 1980. Siamang, lar and agile gibbons. pp. 103-165. In Chivers (1980).

Gron KJ. (2009). Primate Factsheets: Slow loris (*Nycticebus*) Taxonomy, Morphology, & Ecology. Available: http://pin. primate. wisc. edu/ factsheets/entry/slow_loris/taxon [2013, December 1]

Groves, C. P. 2001. *Primate Taxonomy.* Washington, DC: Smithsonian Institution Press. ISBN 978-1-56098-872-4.

Groves, C. P. 2005. Nyticebus menagensis. In Wilson, D. E. ; Reeder, D. M. *Mammal Species of the World: A Taxonomic and Geographic Reference* (3rd ed.). Johns Hopkins University Press., hlm. 111-184. ISBN 978-0-8018-8221-0. OCLC 62265494.

Groves C, and M. Shekelle. 2010. The genera and species of tarsiidae. *International Journal of Primatology,* 31 (6):1071- 1082.

Gurmaya, K. J. 1986. The ecology and behavior of *Presbytis thomasi* in northern Sumatra. *Primates,* 27:151-172.

Gurmaya, K. J. , I. M. W. Adiputra, A. B. Saryatiman, S. N. Danardono, and T. T. H. Sibuea. 1994. A Preliminary study on ecology and conservation of the Java primates in Ujung Kulon Nationak Park, West Java, Indonesia, pp. 87-92. In: Thiery, Anderson, Roeder, Herrenschmidt

Gursky, S. L. 1994. Infant care in the spectral tarsier (*Tarsius spectrum*), Sulawesi, Indonesia. *Int. J. Primatol.,* 15:843 853.

Gursky, S. L. 1995, Group size and composition in the spectral tarsier: implications for social organization. *Trop. Biodiv.,* 3:57 62.

Gursky, S. L. 1997. Modeling Maternal Time Budget: The Impact of Lactation and Infant Transport on the Time Budget of the Spectral Tarsier, *Tarsius spectrum*. PhD dissertation, State University of New York, Stony Brook, NY.

Gursky, S. L. 1998a. The conservation status of the spectral tarsier, *Tarsius spectrum*, in Sulawesi Indonesia. *Folia Primatol,* 69:191 203.

Gursky, S. L. 1998c. The conservation status of two Sulawesian tarsier species: *Tarsius spectrum* and *Tarsius dianae*. *Primate Conserv.,* (18):88 91.

Gursky, S. L. 2000a. Allo-parental care in spectral tarsiers. *Folia Primatol,* 71:39 54.

Gursky, S. L. 2000b. Sociality in the spectral tarsier. *Am. J. Primatol.,* 51:89 101.

Gursky, S. L. 2002a. The determinants of gregariousness in the spectral tarsier. *J. Zool., Lond.,* 256:1 10.

Gursky, S. L. 2002b. Predation on a spectral tarsier by a snake. *Folia Primatol,* 73:60 62.

Gursky, S. L. 2002c. The behavioral ecology of the spectral tarsier. Evol. Anthropol. 11:226 234.

Gursky, S. L. 2003. The effect of moonlight on the behavior of a nocturnal prosimian primate. *Int. J. Primatol.,* 24:351 367.

Gursky-Doyen, S. L. 2010. The function of scent marking of spectral tarsier, pp. 359-369. In: Gursky-Doyen & J. Supriatna.

Gusrky-Doyen, S. L. & J. Supriatna. (eds). 2010. *Indonesian primates.* Springer, New York.

Hart, D. 2007. Predation on primates: a biogeographical analysis. In: Gursky, S. L. and K. A. I. Nekaris (eds.) *Primate Anti-Predator Strategies.* New York, Springer, pp. 27–59.

Husson, S. J. , S. A. Wich, A. J. Marshall, R. A. Dennis, M. Ancrenaz, R. Brassey, M. Gumal, M. , A. J. Hearn, E. Meijaard, T. Simorangkir, *et al.,* 2009. Orangutan distribution, density, abundance and impacts of disturbance. In: Wich, S. A. , Atmoko, S. U. , Setia, T. M. , and van Schaik, C. P. (eds.) *Orangutans: Geographic Variation in Behavioral Ecology and Conservation.* Oxford, UK, Oxford University Press, pp. 77-96.

Indrawan, M. , D. Supriyadi, J. Supriatna, and N. Andayani. 1995. Javan gibbon surviving at mined forest in Gunung Pongkor, Mount Halimun National Park, west Java: conserderable toleration to disturbances. *Asia Primates,* 5(3-4):11-13.

Ishida, H. , E. Hirasaki, and S. Matano. 1992. Locomotion of the slow loris between discountinuous substrates, pp. 139-152 in: S. Matano, R. H. Turtle, & H. Ishida. eds. *Topics in primatology* vol 3. Evolutionary Biology, Reproductive endocrinology and Virology, University of Tokyo Press, Tokyo.

IUCN 2013. The IUCN Red List of Threatened Species. Version 2013. 2. <http://www. iucnredlist. org>. Downloaded on 9 March 2014.

Kappeler, M. 1981. The Javan Silvery Gibbon (*Hylobates lar moloch*): Habitat, Distribution and Number. Ph. D. Thesis, University of Basel, Bsel.

Kitchener, A. C. and C. P. Groves. 2002. New insight of the taxonomy of *Macaca pagensis* of the Mentawai Islands, Sumatra. *Mammalia,* 66:533-542.

Kohlass, A. 1993. Behavior and Ecology of *Macaca nigrescens:* Behaviroal and Social Responses to the Environmental and Fruit availability. Ph. D. Dissertation, University of Colorado, Boulder, Colorado.

Lee, R. J. 1997. The impact of hunting and habitat disturbance on the population dyunamics and behavioral ecology of the crested black macaque (*Macaca nigra*). Ph. D. dissertation, University of Oregon, Eugene

Leighton, D. R. 1987. Gibbons territoriality and monogamy, pp. 135-145. In: Smuts *et al.,* (1987).

Lethinen, J. , K. A. I. Nekaris, V. Nijman, C. N. Zoe Coudrat, and W. Wirdateti. 2013. Distribution of the Javan Slow Loris (*Nyticebus javanicus*): Assesing the presence on East Java, Indonesia. *Folia Primatol.,* 84(3-5):295-295.

Lhota S, B. Loken, S. Spehar, E. Fell, A. Pospech, and N. Kasyanto. 2012. Discovery of Miller's grizzled langur (*Presbytis hosei canicrus*) in Wehea forest confirms the continued existence and extends known geographical range of an endangered primate. *Am. J. Primatol.*, 74:193–198

Madani, G. and K. I. A. Nekaris. 2014. Anaphylactic shock following the bite of a wild Kayan slow loris (Nycticebus kayan): implications for slow loris conservation. *Journal of venomous animals and toxins including tropical diseases,* 20(1):43.

MacKinnon J. and K. MacKinnon. 1980. The behavior of wild spectral tarsiers. *Int. J. Primatol.*, 1(4):361–379.

MacKinnon, J. K. 1974. *In Search of the Red Ape.* London, Collins.

Mansyur, F. I. 2012. Karakteristik Habitat dan Populasi Tarsius (*Tarsius fuscus* Fischer 1804) di Resort Balocci Taman Nasional Bantimurung Bulusaraung Sulawesi Selatan. [Skripsi]. Fakultas Kehutanan, Institut Pertanian Bogor.

Manullang, B. O. , J. Supriatna & D. S. Hadi. 1984. Survey of the Silver Leaf Monkey at Tanjung Karawang Mangrove Forest. West Java. *Proc. Eco. Mangrove,* 2:238 242, Program Man and Biosphere UNESCO, Jakarta.

Markham, R. and C. P. Groves, 1990. Brief communication: weights of wild orangutans. *American Journal of Physical Anthropology,* 81:1-3.

Marshall, A. J. , R. Lacy, M. Ancrenaz, O. Byers, S. Husson, M. Leighton, E. Meijaard, N. Rosen, I. Singleton, and S. Stephens. 2009b. Orangutan population biology, life history, and conservation. Perspectives from population viability analysis models, pp. 311-26. In: Wich, S. , Atmoko, S. U. , Mitra Setia, T. , and van Schaik, C. P. (eds.) *Orangutans: Geographic Variation in Behavioral Ecology and Conservation.* Oxford, UK, Oxford University Press.

Marshall, J. T. and J. Sugardjito. 1986. Gibbons Systematics, pp. 137-186. In: Swindler and Erwin (1986).

Maryanto, I and M. Yani. 2004. The third record of pygmy tarsier (*Tarsius pumilus*) from Lore Lindu National Park, Central Sulawesi, Indonesia. *Trop. Biodiv.*, 8(2) 79-85.

Maryanto, I. , I. Mansyur, D. Sayuthi and J. Supriatna 1997. Morphological variation in the ebony and silver leaf monkeys [*Trachypithecus auratus* (E. Geoffroy 1812) and *Trachypithecus cristatus* (Rafles 1821) from Southeast Asia. *Treubia,* 31:113-131.

Mather, R. 1992. Field study of hybrid gibbons in central Kalimantan. Ph. D. Thesis, Cambridge University, Cambridge, UK.

Megantara, E. N. 1989. Ecology, behavior and sociality of *Presbytis femoralis* in East central Sumatra. *Comp. Primatol. Mon.,* 2:171-301.

Meijaard, E. , G. Albar, Y. Rayadin, Nardiyono, M. Ancrenaz, and S. Spehar. 2010a. Unexpected ecological resil- ience in Bornean Orangutans and implications for pulp and paper plantation management. PLoS One 5, e12813.

Meijaard, E. , D. Buchori, Y. Hadiprakoso, S. S. Utami Atmoko, A. Tjiu, D. Prasetyo, Nardiyono, L. Christie, M. Ancrenaz, and F. Abadi. 2011. Quantifying killing of orangutans and human-orangutan conflict in Kalimantan, Indonesia. PLoS One 6, e27491.

Meijaard, E. , A. Welsh, M. Ancrenaz, S. Wich, V. Nijman, and A. J. Marshall. 2010b. Declining orangutan encounter rates from Wallace to the present suggest the species was once more abundant. PLoS One 5, e12042.

Meijaard, E. , S. Wich, M. Ancrenaz, and A. J. Marshall. 2012. Not by science alone: why orangutan conservationists must think outside the box. *Annals of the New York Academy of Sciences,* 1249:29-44.

Meijard, E. and V. Nijman. 2000. Distribution and Conservation of proboscis monkey (*Nasalis larvatus*) in Kalimantan, Indonesia. *Biol Conserv.,* 92:15-24.

Melisch, R. & I. W. A. Dirgayusa. 1996. Notes on the grizzled leaf monkey (*Presbytis comata*) from two nature re- serves in the West Java, Indonesia. *Asian Primates,* 6(2):5-11.

Merker, S. 2003 Endangered or adaptable? - Dian's tarsier *Tarsius dianae* in Sulawesi's rain forests. Georg-August University, Gottingen.

Merker, S. , I. Yustian, and M. Muehlenberg. 2005. Responding to forest degradation: altered habitat use by Dian's tarsier *Tarsius dianae* in Sulawesi, Indonesia. *Oryx,* 39:189-195.

Merker, S. , & C. P. Groves. 2006. *Tarsius lariang:* a new primate species from western central Sulawesi. *Int. J. Primatol.,* 27(2):465-485

Merker, S. and M. Muehlenberg. 2000. Traditional land use and tarsier-human influences on population densities of *Tarsius dianae*. *Folia primatol.*, 71:426-428.

Merker, S. and I. Yustian. 2008. Habitat use analysis of Dian's tarsier (Tarsius diana) in a mixed species plantation in Sulawesi, Indonesia. *Primates*, 49:161-164

Merker, S. 2010. "The population ecology of Dian's tarsier, pp. 371-382". In: Indonesia Primates; Developments in Primatology: Progress and Prospects (Gursky-Doyen & Supriatna), Chicago: Springer

Merker S, C. Driller C. H. Dahruddin, Wirdateti, W. Sinaga, D. Perwitasari-Farajallah & M. Shekelle. 2010. Tarsius wallacei: A New Tarsier Species from Central Sulawesi Occupies a Discontinuous Range". Int. J. Primatol: DOI 10. 1007/s10764-010-9452-0

Merker, S. and C. P. Groves. 2006. Tarsius lariang: A new primate species from western central Sulawesi. *Int. J. Primatol.*, 27:465-485.

Merker, S. , C. Driller, D. Perwitasari-Farajallah, J. Pamungkas and H. Zischler. 2009. Elucidating geological and biological processes underlying the diversification of Sulawesi tarsiers. Proc. Natl. Acad. Sci. USA 106:8459-8464.

Merker, S. , C. Driller, H. Dahruddin, Wirdateti, W. Sinaga, D. Perwitasari-Farajallah and M. Shekelle. 2010. Tarsius wallacei: a new tarsier species from Central Sulawesi occupies a discontinuous range. *Int. J. Primatol.*, 31:1107-1122.

Miller Jr. , G. S. and N. Hollister . 1921. Twenty new mammals collected by H. C. Raven in Celebes. *Proc. Biol. Soc. Wash.*, 34:93-104.

Mitani, J. 1990. Demographic of agile gibbons (Hylobates agilis). *Int. J. Primatol.*, 11:411-425.

Mittermeier R. A. , J. Ratsimbazafy, A. B. Rylands, L. Williamson, J. Oates. D. Mbora, J. U. Ganzhorn, E. Rodríguez-Luna, E. Palacios, E. W. Heymann, M. C. Kierulff, Y. Long, J. Supriatna, C. Roos, S. Walker, S. and J. M. Aguiar 2007. Primates in Peril: The world's 25 most endangered primates in 2006-2008. *Primate Conservation*, 22:1-40.

Mittermeier, R. A. , A. B. Rylands, & D. E. Wilson, eds. 2013. Handbook of the Mammals of the World. Vol. 3. Primates. Barcelona: Lynx Edicions

Moore, R. S. , K. A. I. Nekaris & S. Wihermanto. 2011. Primate polination and the origins of prehensile. Poster in Primate Society of Great Britain Conference in 2011.

Munds, R. A. , K. A. I. Nekaris, and S. M. Ford. 2013. Taxonomy of the Bornean slow loris, with new species *Nycticebus kayan* (Primates, Lorisidae). *American Journal of Primatology, 75 (1):46–56.* doi:10. 1002/ajp. 22071. PMID 23255350.

Musser, G. G and M. Dagosto. 1987. "The identity of *Tarsius pumilus.* A pygmy species endemic to the montane mossy forests of central Sulawesi". *Am. Mus. Novit.* 2867:1-53.

Mustari, A. H. , F. I. Mansyur, and D. Rinaldi. 2013. Karakteristik habitat dan populasi tarsius (Tarsius fuscus Fischer 1804) di resort Balocci, Taman Nasional Bantimurung Bukusaraing, Sulawesi Selatan). *Media Konservasi,* 18 (1):47-53.

Napier, J. R. and P. H. Napier 1967. *A Handbook of living primates.* Academic Press, London.

Nater, A., M. P. Mattle-Greminger,. A. Nurcahyo, M. G. Nowak, M. de Manuel, T. Desai, C. Groves, M. Pybus, T. B. Sonay, C. Roos. 2017. Morphometric, behavioral, and genomic evidence for a new orangutan species. *Curr. Biol.,* 27, 3487-3498.

Nekaris, KAI and S. Jaffe. 2007. "Unexpected diversity of slow lorises (Nycticebus spp) in the Javan pet trade: implications for slow lorises taxonomy". *Contrib Zoology,* 76:187-196

Nekaris, K. A. I. , G. V. Blackham, and V. Nijman 2008. Implications of low encounter rates in five nocturnal species (Nycticebus spp). Biodiversity Conservation 17 (4):733-747.

Nekaris, K. A. I. and R. Munds. 2010. "Chapter 22: Using facial markings to unmask diversity: the slow lorises (Primates: Lorisidae: Nycticebus spp.) of Indonesia. In Gursky-Doyen, S. ; Supriatna, J. *Indonesian Primates.* New York: Springer. pp. *383– 396.* doi:10. 1007/978-1-4419-1560-3_22. ISBN 978-1-4419-1559-7.

Nekaris, K. A. I. , R. S. Moore, E. J. Rode, B. G. Fry. 2013. Mad, bad and dangerous to know: the biochemistry, ecology and evolution of slow loris venom. *Journal of Venomous Animals and Toxins including Tropical Diseases,* 19 (1):21

Nekaris, K. A. I. , I. Poindexter, K. D. Reinhardt, M. Sigaud, F. Cabana, W. Wirdateti, and V. Nijman. 2017. Coexistence between Javan Slow Lorises (*Nycticebus javanicus*) and Humans in a Dynamic Agroforestry Landscape in West Java, Indonesia. *Int. J. Primatol.,* 38 (2):303-320.

Niemitz, C. 1984. Vocal communication of two tarsier species (*Tarsius bancanus* and *Tarsius spectrum*). In: *Biology of Tarsiers,* C. Niemitz (ed.) , pp. 129-142. Gustav Fischer Verlag, New York.

Niemitz, C. , A. Nietsch , S. Warter and Y. Rumpler. 1991. *Tarsius dianae:* a new primate species from Central Sulawesi (Indonesia). *Folia Primatol.,* 56:105-116.

. Nietsch, A. 1999. Duet vocalizations among different populations of Sulawesi tarsiers. *Int. J. Primatol.,* 20:567 583.

Nietsch, A. and M. L. Kopp. 1998. Role of vocalization in species differentiation of Sulawesi tarsiers. *Folia Primatol.,* 68 (suppl. 1):371 378.

Nietsch, A. and C. Niemitz. 1993. Diversity of Sulawesi tarsiers. Z. Saugetierkd. 67:45-46.

Nijman, V. and Sozer 1995. Recent observation of the grizzled leaf monkeys (*Presbytis comata*) and extension of Javan gibbon (*Hylobates moloch*) in Central Java. *Trop Biodiv.,* 3:45-48

Nijman, V. and J. Supriatna 2008. *Trachypithecus auratus.* In: The IUCN Redlist of threaten species. Version 2011, url:http//www. iucnredlist. org/details/22034.

Nijman, V. 1997 Geographical variation in pelage characteristic in *Presbytis comata* Desmarest, 1822, Mammalia, Primates, Cercopithecidae). Z. Saugeter 62:257-264.

Nijman, V and B. van Balen 1998 Faunal survey of the Dieng Mountain , Central Java, Indonesia: distribution and endemic primate taxa. Oryx 32:145-156.

Nijman, V. 2004. Effects of disturbance habitat and hunting n densities and biomass of the endemic hose's leaf monkey *Presbytis hosei* (Thomas1889) (Mammalia: Primates: Cercopithecidae) in East Borneo. Contrib Zool., 73:283-291.

Nijman, V. 2005. Rapid decline of Hose's langur in Kayan Mentarang National Park. *Oryx.,* 39(2):223-226.

Nijman, V. and M. Lammerlink 2008. *Presbytis natunae.* In IUCN 2008. 2008 IUCN Red List of threatened species Version 12. 1. www. iucnredlist. org.

Nijman, J. , E. Meijard, and J. Hon 2008. *Presbytis hosei.* In IUCN 2008. 2008 IUCN Red List of threatened species. Version 12. 1. www. iucnredlist. org.

Nijman, V. 2010. Ecology and Conservation of the Hose's langur group (Colobinae: *Presbytis hosei, P. canicrus, P. sabana*), pp. 269-284. In: *Indonesia Primates; Developments in Primatology: Progress and Prospects* (Gursky-Doyen & Supriatna). Chicago: Springer

Nowak, R. M. 1999. *Walker's Primates of the World.* Baltimore: The Johns Hopkins University Press.

Oates, J. F. , A. G. Davies, and E. Delson 1994. The diversity of Living colobines, pp. 45-73. In David and Oates. *Colobine Monkeys: their ecology, behaviour and evolution,* eds., Cambridge: Cambridge University Press.

Oi, T. 1990. Population organizations of wild pig-tailed macaques in West Sumatra. *Primates,* 31:15-31

Oi, T. 1996. Sexual behavior and mating system of wild pig-tailed macaques in West Sumatra, pp. 342-368. In: Fa and Lindburg.

Paciulli, L. M. 2004. The effects of logging, hunting, and vegetation on densities of the Pagai Mentawai Island primates (Indonesia). Dissertation Thesis, State University of New York, Stony Brook.

Paciulli, L. M. 2010. The relationship between non human primate densities and vegetation of the Pagai, Mentawai Island, Indonesia, pp. 199-215. In: *Indonesia Primates; Developments in Primatology: Progress and Prospects* (Gursky-Doyen & Supriatna). Chicago: Springer

Payne, J. , C. M. Francis, and K. Phillips 1985. *A Field Guide to the mammals of Borneo.* Sabah Society and WWF Malaysia, Kota Kinabalu, Malaysia.

Perbatakusuma, E. A. , Onrizal, Ismail, H. Soedjito, J. Supriatna and H. Wijayanto 2008. Struktur Vegetasi dan Simpanan Karbon Hutan Hujan Primer di Batang Toru Indonesia (Vegetation Structure and

Carbon Stock at the Batang Toru Tropical Rain Forest, Indonesia). *Jurnal Biologi Indonesia,* 5 (2):187-199.

Pournelle, G. H. 1967. Observations on reproductive behaviour and early postnatal development of the proboscis monkey Nasalis larvatus orientalis. *International Zoo Yearbook,* 7:90-92.

Prasetyo, D. S. S. Utami and J. Supriatna 2012. Nest Structures in Bornean Orangutans. *Jurnal Biologi Indonesia,* 8(2):217-227.

Putri, P. R. , E. J. Rode, N. L. Winarni, Wirdateti, and K. A. I. Nekaris. 2013. Habitat and substrate use by the Javan slow loris (Nycticebus javanicus) in a Talun plantation, Cipaganti, West Java, Indonesia. Poster presentation at the spring meeting 2013 of the Primate Society of Great Britain

Putri, P. R. Aktivitas Harian dan Penggunaan Habitat Kukang Jawa (*Nycticebus javanicus*) di Talun Desa Cipaganti, Garut, Jawa Barat. Skripsi Fakultas Matematika Dan Ilmu Pengetahuan Alam, Departemen Biologi, Depok.

O'Brien, T. G. , M. F. Kinnard, A. Nurcahyo, M. Prasetyaningrum, M. Iqbal 2003. Fire, demography and persistence of Siamang (*Symphalangus syndactylus*, Hylobatidae) in Sumatran Rain Forest. *Animal Conserv.,* 6:115-121.

Rao, M. and van Schaik, CP. 1997. The behavioral ecology of Sumatran orangutan in logged and unlogged forest. *Trop Biodiv.,* 4(2):173-185.

Richardson, M., R. A. Mittermeier, A. B. Rylands & Konstant, B. 2008. *Macaca nemestrina.* The IUCN Red List of Threatened Species 2008:. Downloaded on 24 October 2016.

Rijksen, H. D. 1978. A Field study of orangutan (*Pongo pygmaeus abelii* Lesson 1827): Ecology, Behaviour and Conservation.

Rijksen, H. D. and E. Meijaard. 1999. *Our Vanishing Relative. The Status of Wild Orang-Utans at the Close of the Twentieth Century.* Dordrecht, The Netherlands, Kluwer Academic Publishers.

1. Riley, E. P. 2008. Ranging patterns and habitat use of Sulawesi Tonkean macaques (Macaca tonkeana) in a human modified habitat. *Am. J. Primatol.,* 70:670-679.

Riley, E. P. 2005. The loud call of the Sulawesi Tonkean macaque, Macaca tonkeana. *Trop Biodiv.,* 8(3):199-209.

Riley, E. P. , B. Suryobroto, and D. Maestripieeri. 2007. Distribution of Macaca ochreata and Identification of Mixed Ochreata-Tonkeana Groups in South Sulawesi, Indonesia. *Primate conservation,* 22(1):129-133.

Richter, C. , A. Taufiq, K. Hodges, J. Ostner, and O. Schulke. 2013. *Ecology of Endemic Primate Species (Macaca siberu) on Siberut Island, Indonesia.* SpringerPlus 2:137.

Rode-Margono, J. E. Rademaker, M. Wirdateti, Strijkstra, A. , Nekaris, K. A. I. 2014. Noxious arthropods as potential prey of the venomous Javan slow loris (Nycticebus javanicus) in a West Javan volcanic agricultural system. *Journal of natural history,* 49 (31-32):1949-1959.

Roos, C. , R. Boonratana, J. Supriatna, J. R. Fellowes, C. P. Groves, & D. Stephen. 2014. An Updated Taxonomy and Conservation Status Review of Asian Primates. *Asian Primate Journal,* 4 (1):2-38.

Rosembaum, B. , T. O'Briens, M. Kinnard, J. Supriatna 1998. Population Densities of Sulawesi Crested Black Macaques (Macaca nigra) on Bacan and Sulawesi, Indonesia: Effects of Habitat Disturbance and Hunting. *Am. J. Primatol.,* 44:89-106.

Rowe, N. 1996. *The Pictorial Guide to the Living Primates.* Pogonias Press, East Hampton, NY. Google Scholar

Ruhiyat, Y. 1983. Socioecological Study of *Presbytis aygula* in West Java. *Primates,* 24:344-359.

Ruhiyat, Y. 1986. Preliminary study of proboscis monkey (Nasalis larvatus) in Gunung Palung Nature Reserve, West Kalimantan. Pp. 59-69. In: Kyoto University Overseas Reseaerch Reporrt of Studies in Asian non-human primates. No. 5, Kyoto University, Primate Research Institute, Kyoto.

Salter, R. E. , N. A. Mackenzie, N. Nightingale, K. M. Aken, and P. K. Chai. 1985. Habitat use, ranging behaviorand food habits of proboscis monkey (*Nasalis larvatus* Wurmb)in Serawak. *Primates,* 26 (4):436-451.

Setiawan, A., T. S. Nugroho, Djuwantoko, & S. Pudyatmiko. 2009. A Survei of *Presbytis hosei canicrus* in East Kalimantan, Indonesia. *Primate Conservation,* 24:1-5.

Shagir K. J. , I. A. Rasjid, P. Wibowo, S. Fajrin and Paisal. 2011. Identifikasi dan Pemetaan Sebaran Tarsius Pada Kawasan Taman Nasional

Bantimurung Bulusaraung di Kabupaten Maros. Bantimurung: Balai Taman Nasional Bantimurung Bulusaraung.

Setiawan A, T. S. Nugroho, Y. Wibisono, V. Ikawati, J. Sugardjito. 2012. Population density and distribution of javan gibbon (*Hylobates moloch*) in Central Java, Indonesia. Biodiversitas 13(1):23-27. https://doi. org/10. 13057/ biodiv/d130105.

Shekelle, M. 2003. Taxonomy and biogeography of Eastern Tarsiers. PhD. Dissertation. Saint Louis, USA Washington University.

Shekelle, M. , S. M. Leksono, L. S. I. Ichwan, and Y. Masala. 1997. The natural history of the tarsiers of north and central Sulawesi. *Sulawesi Primate Newsl.,* 4(2):4 11.

Shekelle M. dan S. M. Leksono. 2004. Rencana Konservasi di Pulau Sulawesi dengan Menggunakan Tarsius sebagai Flagship Spesies. *Biota* 9(1):1-10.

Shekelle, M. & A. Salim. 2008. *Tarsius pumilus.* The IUCN Red List of Threatened Species 2008:e. T21490A9288636. http://dx. doi. org/10. 2305/IUCN. UK. 2008. RLTS. T21490A9288636. en. Downloaded on 06 April 2019.

Shekelle, M. 2008. Distribution and biogeography of tarsiers. In: Primates of the Oriental Night , M. Shekelle , I. Maryanto , C. P. Groves , H. Schulze and H. Fitch-Snyder (eds.), pp. 13-28. Indonesian Institute of Sciences. Cibinong, Indonesia.

Shekelle M, C. P. Groves, S. Merker , and J. Supriatna. 2008. *Tarsius tumpara:* A New Tarsier Species from Siau Island, North Sulawesi. *Primate Conservation,* (23):55-64.

Shekelle , M. , C. P. Groves, S. Gursky, I. Neri-Arboleda, and A. Nietsch. 2008. A method for multivariate analysis and classification of tarsier tail tufts. In: *Primates of the Oriental Night,* M. Shekelle, I. Maryanto, C. P. Groves, H. Schulze and H. Fitch-Snyder (eds.), pp. 71-84. Indonesian Institute of Sciences. Cibinong, Indonesia.

Shekelle, M. and A. Salim. 2009. Acute conservation threat to two tarsier species in Sangihe island chain, North Sulawesi, Indonesia. *Oryx,* 43 (3):419-426.

Shekelle, M. , C. P. Groves, I. Maryanto, and R. Mittermeier. 2017. Two New Tarsier Species (Tarsiidae, Primates) and the Biogeography of Sulawesi, Indonesia. *Primat Conserv.,* 31:1-9.

Shekelle, M. 2008. Distribution of tarsier acoustic forms, north and central Sulawesi: with notes on the primary taxonomy of Sulawesi's tarsiers. In: *Primates of the Oriental Night,* M. Shekelle, C. P. Groves, I. Maryanto, H. Schulze and H. Fitch-Snyder (eds.), pp. 35 50. Research Center for Biology, Indonesian Institute of Sciences, Bogor, Indonesia.

Shekelle, M. and S. M. Leksono. 2004. Rencana konservasi di Pulau Sulawesi: dengan menggunakan Tarsius sebagai flagship spesies. *Biota,* 9 (1):1 10.

Shekelle, M. , S. Leksono, L. S. I. Ichwan and Y. Masala. 1997. The natural history of the tarsiers of north and central Sulawesi. *Sulawesi Primate Newsl.,* 4(2):4 11.

Shekelle M. , J. C. Morales, C. Niemitz, L. S. I. Ichwan and D. M. Melnick. 2008. The distribution of tarsier mtDNA haplotypes for parts of north and central Sulawesi: a preliminary analysis. In: *Primates of the Oriental Night,* M. Shekelle, C. P. Groves, I. Maryanto, H. Schulze and H. Fitch-Snyder (eds.), pp. 51–69. Research Center for Biol ogy, Indonesian Institute of Sciences, Bogor, Indonesia.

Shekelle, M. , R. Meier, I. Wahyu, Wirdateti and N. Ting. 2010. Molecular phylogenetics and chronometrics of Tarsiidae based on 12s mtDNA haplotypes: evidence for Miocene origins of crown tarsiers and numerous species within the Sulawesian clade. *Int. J. Primatol.,* 31:1083 1106.

Sinaga, W. , Wirdateti, E. Iskandar and J. Pamungkas J. 2009. Pengamatan habitat pakan dan sarang Tarsius (*Tarsius sp.*) wilayah sebaran di Sulawesi Selatan dan Gorontalo. *Jurnal Primatologi Indonesia,* 6 (2):41-47.

Singleton, I and C. van Schaik. 2001. Orangutan home range and its determinants in a Sumatran swamp forest. *Int. J. Primatol.,* 22 (6):877-911.

Singleton, I. , S. Wich, S. Husson, S. Stephen, S. S. Utami-atmoko, M. Leighton, N. Rossen, K. Taylot-Holzer, R. Lacy, O. Byers. 2004.

Orangutan Population and Habitat Viability Assesment. Final Report IUCN SSC. Captive Breeding Specialist Group, Apple Valley, Minnesota.

Sloan, S, J. Supriatna, M. J. Campbell, M. Alamgir, and W. F. Laurance. 2018. Newly discovered orangutan species requires urgent habitat protection. *Current Biology,* 28:R1-R3, June 4, 2018.

Steenbeek, R. , P. Assink, and S. A. Winh. 1999. Tenure related changes in wild Thomas's langurs: Loud calls. *Behaviour,* 136:627-650.

Sugardjito, J. C. H. Southwick, J. Supriatna, A. Kohlhaas, and N. Lerche. 1989. Population survey of macaques in Northern Sulawesi. Am. J. Primatol 18 (4): DOI 10. 1002/ajp. 1350180403.

Sugardjito, J. and A. Adhikerana. 2010. Measuring performance of orangutan protection and monitoring unit: implication for species conservation. Pp9-22. In: S. Gursky-Doyen and J. Supriatna (eds). *Indonesia primates.*

Supartono, T. L. B. Prasetyo, A. Hikmat, and A. P. Kartono. 2016. Response of Group size to edge and Population density of grizzled leaf monkey (*Presbytis comata*) in lowland and hills forest of Kuningan district. *Zoo Indonesia,* 25(2):107-121.

Supriatna, J. Asri A. Dwiyahreni, Nurul Winarni2, Sri Mariati and Chris Margules 2017. Deforestation of Primate Habitat on Sumatra and Adjacent Islands, Indonesia. *Primate Conservation,* 31(3):. 71-82.

Supriatna, J. , N. Winarni, A. A. , Dwiyahreni 2015. Primates of Sulawesi: An Update on Habitat Distribution, Population and Conservation. *Taprobanica,* 7(3):170-192.

Supriatna, J. I. Wijayanto, B. O. Manullang, D. Anggraeni, Wiratno, S. Ellis 2002. The state of siege for Sumatra's forest and protected areas: Stakeholders view during devolution, and political plus economic crises in Indonesia. Proc. IUCN/WCPA-East Asia, pp. 439-456, Taipei, Taiwan.

Supriatna, J. A. Yanuar, Martarinza, H. T. Wibisono, R. Sinaga, I. Sidik and S. Iskandar 1996. A Preliminary survey of longtailed and pigtailed macaques (*Macaca fascicularis* and *M. nemestrina*) in Lampung, Bengkulu and Jambi provinces, Southern Sumatra, Indonesia. *Tropical Biodiversity,* 3 (2):131-140.

Supriatna, J. and E. Hendras. 2000. *Panduan lapangan Primata Indonesia* (Indonesia's Primate Field Guide). Yayasan Obor Jakarta, 332 p.

Supriatna, J. and B. O. Manullangs (ed). 1999. Proceeding of the International Workshop on Javan Gibbon Rescue and Rehabilitation. Center for Biodiversity and Conservation Studies, University of Indonesia, Depok.

Supriatna, J. , B. O. , Manullang and E. Soekara 1986. Group Composition, Home range and Diet of the Maroon Leaf monkey (*Presbytis rubicunda*) at Tanjung Puting nature Reserve, Central Kalimantan, Indonesia. *Primates,* 27(2):185190.

Supriatna, J. , E. Perbatakusuma, A. H. Damanik, H. Hasbullah & A. Ario 2014. Sumatran orangutan as a flagship for conserving biodiversity and parks: Lesson learnt from North Sumatra Conservation Awareness Programmes. *Asia Primate Journal,* 4(2):52-59.

Supriatna, J. 2000. Status Konservasi Satwa Primata. Dalam: Konservasi Satwa Primata: Tinjauan Ekologi, Sosial Ekonomi, dan Medis dalam Pengembangan Ilmu Pengetahuan dan Teknologi (P. Yuda &S. I. O. Salasia eds), Fakultas Kedokteran Hewan dan Kehutanan Univ. Gadjah Mada, Yogyakarta, p. 3-12.

Supriatna, J. , C. Adimuntja, T. Mitrasetia, E. Willy, D. Rufendi and B. O. Manullang. 1989. Chemical Analysis of Foodplant parts of two Sympatric monkeys (Presbytis auratus and Macaca fascicularis). *Biotrop Special Publication,* 37:161169

Supriatna, J. , J. W. Froehlich, J. M. Erwin & C. H. Southwick. 1992. Population, Habitat and Conservation of Macaca maurus, M. tonkeana and their putative hybrids. *Tropical Biodiversity,* 1(1):31-48.

Supriatna, J. , K. G. Gurmaya, Wahyudi, W. A. . Sriyanto, R. Tilson & U. Seal. 1994. *Javan Gibbon and Langur; Population Habitat Viable Analysis.* CBSG-IUCN, Minnesota.

Suzuki, A. 1984. The distribution of primates and survey on the affection of forest fires, 1983, in and around Kutai nature reserve of East Kalimantan. Kyoto University Overseas Res. Rep Stud Asian Non Human Primates, 3:53-58.

Tenaza, R. 1987. The status of primates and their habitat in Pagai islands, Indonesia. *Primate Conserv.,* 8:104-110.

Thorn, J. S. , V. Nijman, D. Smith, and K. A. I. Nekaris. 2009. Ecological niche modelling as a technique for assessing threats and setting conservation priorities for Asian slow lorises (Primates, *Nycticebus*). *Divers Distrib,* 15:289-298.

Tilson, R. L. 1976. Infant coloration and taxonomic affinity of the Mentawai island leaf monkey, Presbytis potenziani. J. Mammal 57:766-769.

Tilson, R. L. , and R. Tenaza. 1976. Monogamy and duetting on old World monkeys. Nature 263:320-321.

Tilson, R. L. , and R. Tenaza. 1982. Interspecific spacing between gibbons (*Hylobates klossi*) and langurs (*Presbytis potenziani*) on Siberut Island, Indonesia. *Am. J. Primatol,* 2:355-361.

Tremble, M. , Y. Muskita and J. Supriatna. 1993. Field observation of *Tarsius diana* at Lore Lindu National Park. , Central Sulawesi, Indonesia. *Tropical Biodiversity,* 1(2):67-76.

Ungar, P. S. 1995. Fruit preferences of four sympatric primate species at Ketambe, Northern Sumatra, Indonesia. *Int. J. Primatol,* 16:221-245

Ungar, P. S. 1995. Fruit preference of four sympatric primates at Ketambe, Northern Sumatra, Indonesia. *Int. J. Primatol,* 16:221-245.

Utami, S. S, and van Hoff, JARAM 1997. Meat eating by adult female Sumatran orangutan (Pongo pygmaeus abilii). *Am. J. Primatol* 43:156-165.

Utami, S. S. , B. Goosens, M. W. Bruford, J. de Ruiter, van Hoof. JARAM 2002. Male bimaturism and reproductive success in Sumatran orangutans. *Behav. Ecol.,* 13:643-652.

Utami, S. S. and C. van Schaik 2010. The natural history of Sumatran primates. Pp. 41-55. In: S. Gursky-Doyen and J. Supriatna (eds). *Indonesia Primates.*

van Schaik, C. , M. S. Azwar, and D. Priatna. 1995. Population estimates and habitat preferences of orangutan based line transect nests, pp. 129-147. In: *R. D. Nadler, B. M. F. Galdikas,* L. K. Sheeran, N. Rosen (eds). The neglected ape, Plenum, New York.

van Schaik, C. and M. Horstermann. 1994. Predation risk and the number of adult males in a primate groiup: a comparative test. *Behav. Ecol. Sociobiol,* 35:261-272.

van Schaik, C. 2004. Among orangutan: red apes and the rise of human culture. Cambridge, MA: Harvard University Press.

van Schaik, C. , K. A. Monk, and J. MY. Robertson. 2001. Dramatic decline in orangutan numbers in the Leuser Ecosystem. , Northern Sumatera. Oryx 35:15-25.

Vun, V. F. , M. C. Mahani, M. Lakim, A. Ampeng, & B. M. Md-Zain. 2011. Phylogenetic relationship of leaf monkey (*Presbytis*, Colobinae) based on Cytochrome b abs 12S r RNA genes. *Genet. Mol. Res*, 10(1):368-391.

Watanabe, K. 1981. Variation in group composition and population density of the two sympatric Mentawaian leaf monkeys. *Primates*, 22:145-160.

Waltert, M. , C. Abbeg, zt, Zigler, S. Hadi, D. Priatna, and K. Hodges. 2008. Abundance and community structure of Mentawaian primates in the Paleonan forest, North Siberut, Indonesia. *Oryx*, 42:375-379.

Whitten, A. J. 1982. Diet and feeding behaviour of Kloss gibbon in Siberut island, Indonesia. *Folia Primatol*, 37:177-208.

Wich, S. A. , S. Konski, H. de Vries, C. P. van Scahik 2003. Individual and contextual variation in Thomas male loud calls. *Ethology*, 109(1):1-13.

Wich, S. A. , P. R. Assink, E. H. M. Sterck. 2004. Thomas langur (*Prebystis thomasi*) discriminate between call of young solitary versus older group-living males: factor in avoiding infanticide. *Behavior*, 141:41-51.

Wich, S. A. and E. H. M. Sterck 2010. Thomas langurs: Ecology, Sexual conflict and Social Dynamics. In pp. 285-308. In: *Indonesia Primates;* Developments in Primatology: Progress and Prospects (Gursky-Doyen & Supriatna). Chicago: Springer

Wich, S. A. E. Meijard, A. J. Marshall, S. Husson, M. Acrenaz, R. C. Lacy, C. P. van Schaik, J. Sugardjito, T. Simorangkir, K. Traylor-Holzer, M. Doughty, J. Supriatna, R. Dennis, M. Gumal, C. D. Knott and I. Singleton. 2008. Distribution and conservation status of the orangutan (*Pongo* spp) on Borneo and Sumatera: how many remain? *Oryx*. 42 (3):329-339.

Wich, S. A. , S. S. Utami-Atmoko, T. Mitra Setia, Djojosudarmo, and M. L. Geurts. 2006. Dietary and energetic responses of Pongo abelii to fruits availability fluctuability. Int. J. Primatol 27:1535-1550.

F. 2002. Behavior and ecology of wild slow lorises (Nycticebus coucang): social organization infant care system, and diet. Ph. D. Dissertation, University of Bayreut, Bayreuth, Germany

Weins, F. , A. Zitzmann, and N. A. Hussein. 2006. Fast food for slowlorises: is low metabolism related to secondary compoundsin high-energy plant diet? *J Mammal,* 87:790-798.

Wilson, C. C. and W. L. Wilson. 1976. Behavioral and Morphological variation among primate population in Sumatra. Yearb. *Phys. Anthropol,* 20:207-233.

Wirdateti, and S. Setyorini 2004. Pakan dan habitat kukang (Nycticebus coucang) di hutan lindung perkampungan Baduy, Rangkasbitung-Banten Selatan. *Biodiversitas,* 6 (1):45-49.

Wirdateti, H. Dahrudin, and A. Sumadijaya. 2011. Sebaran habitat kukang Jawa (Nycticebus javanicus) di lahan pertanian (Hutan rtakyat) wilayah Lebak (Banten) dan Gunung Salakj (Jawa Barat). *Zoo Indonesia,* 20(1):

Wirdateti & H. Dahrudin 2008. An exploration on the habitat, feeds and distribution of Tarsius tarsier (tarsier) in Selayar Island and Patunuang Nature Reserve, South Sulawesi. *Biodiversitas,* 9 (2):152-155.

Whittaker, D. J. 2005. New population estimates for the endemic Kloss's gibbon Hylobates klossii on the Mentawai Islands, Indonesia. *Oryx,* 39:458-61.

Whittaker, D. J. 2006. A Conservation action plan for Mentawai primate. *Primate Conservation,* 58(67):95-105.

Woodruff, D. S. , K. Monda, and R. E. Simmons. 2005. Mitichondrial DNA sequence variationand subspecific taxonomy in the white-handed gibbons, *Hylobates lar.* Nat Hist. J. Chulalongkorn Univ. 2005 (supll 1):71-78.

Wolf, K. E. and J. G. Fleagle. 1977. Adult male replacement in a group of silvered leaf monkey (Presbytis cristatus) at uala Selangor, Malaysia. *Primates,* 18:949-955.

Yanuar, A. , D. Bekti, and C. Saleh. 1993. The Status of Karimata primates *Presbytis rubicunda carimatae* and *Macaca fascicularis carimatensis* in Karimata island, Indonesia. *Trop Biodiv.,* 1:157-162.

Yanuar, A. and J. Supriatna. 2018. The Status of Primates in the Southern Mentawai Islands, Indonesia. *Primate Conservation,* 32:11p

Yeager, C. P. 1995. Does intraspecific variation in social systems explain reported differences in the social structure of the proboscis monkey (*Nasalis larvatus*)? *Primates,* 36:577-584.

Yeager, C. P. 1992. Proboscis monkey (*Nasalis larvatus*) social organization: the nature and possible functions of intergroup patterns of association. *Am. J. Primatol.,* 26:133-138.

Yeager, C. P. 1989. Proboscis Monkey (*Nasalis larvatus*) Social Organization and Ecology. Ph. D. thesis, University of California - Davis.

Yeager, C. P. 1990. Proboscis monkey social organization: group structure. *Am. J. Primatol,* 20:95-106

Yeager, C. P. 1991. Possible antipredator behavior associated with river crossings by proboscis monkeys (*Nasalis larvatus*). Am. J. Primatol 24 (1):61-66.

Yi, Y. , M. Shekelle, and. J. C. Choe. 2015. Discriminating tarsier acoustic forms from Sulawesi's northern peninsula with spectral analysis. Paper presented at the 84th Annual Meeting of the American Association of Physical Anthropologists, 25 29 March 2015, St. Louis, MO.

Yustian, I, S. Merker, J. Supriatna, and N. Andayani. 2008. Relative density of *Tarsius dianae* in man-influenced habitats of Lore Lindu National park, Central Sulawesi, Indonesia. Asian Primates Journal 1 (1):10-16.

Index

© The Author(s), under exclusive license to Springer Nature Switzerland AG 2022
J. Supriatna, *Field Guide to the Primates of Indonesia*,
https://doi.org/10.1007/978-3-030-83206-3

About the Author

After finishing his Bachelor of Science (Sarjana) in Biology (1978) from Universitas Nasional Jakarta, Master of Science (1986) and Doctorate degree (1991) from the University of New Mexico, Albuquerque-USA, plus pre and post-doctoral at Columbia University in New York, he serves as Lecturer then Professor at the University of Indonesia. He had been appointed as Country Director then Vice President of the major international organization based in Washington DC, Conservation International or CI (1994-2010). He retired from CI and since January 2011 to date, is serving as Chairman of Research Center for Climate Change and Institute for Sustainable Earth and Resources of the University of Indonesia, co-director of the Association of Pacific Rim University (APRU based in the National University of Singapore) of CMAS (Climate Change Mitigation and Adaptation Strategy) based in the University of San Diego, USA. He also serves as Country chair of United Nations Sustainable Development Solution Network (UN SDSN) for Indonesia. Since 2011 he has been appointed as a member of Indonesia Academy of Science (AIPI). He is also serving as a member of board of Trustee of the Universitas Indonesia 2014-2019, Board of trustee members of UID Foundation, Belantara Foundation, Indonesia Climate Change Trust Fund of BAPPENAS, Bornean Orangutan Society Foundation, CSF Indonesia, and TFCA Kehati (Indonesia Biodiversity Foundation).

232
J. Supriatna, *Field Guide to the Primates of Indonesia*,
https://doi.org/10.1007/978-3-030-83206-3

His research interest is on Primatology, Wildlife Tourism, Landscape conservation, Biodiversity Conservation, Climate change Mitigation and Adaptation and interlinked between plus Environmental policy.

For his dedication to the environment and biodiversity works, he received a distinguished award of the most Excellence Order of Golden Ark from his Royal Highness Prince Bernhard of the Netherland in 1999. In 2009, he also received the most privileged Award from President B.J. Habibie of Indonesia, or Habibie Award for outstanding achievements on research on Natural Science. In 2010, he received Terry MacManus Award of the United States of America for his dedication on conserving nature. In 2011, he received Achmad Bakrie Award on Science for his dedication in developing Field Biology and Conservation in Indonesia. In 2017, he received Lifetime Achievement in the Field of Biodiversity Conservation from Conservation International. IUCN of SSC-Primate Specialist Group named after him, one of new tarsier primate species from Gorontalo, North Sulawesi, *Tarsius supriatnai*. He has served as an editorial board member of several international journals such as IUCN Parks Journal, Asian Primate Journal (IUCN-Primate Specialist Group), Tropical Conservation Science Journal, American Journal of Wildlife Policy and Law, and Biosphere Conservation. He published 21 books mostly on Indonesia's environment and Biodiversity. He published more than 150 papers in reputable International Journals (Science, Nature, Conservation Biology, Evolution, Peer J, Current Biology, Primates, International Journal of Primatology, Primate Conservation, many others).

Printed in the United States
by Baker & Taylor Publisher Services